Stable Numerical Schemes for Fluids, Structures and their Interactions

To my daughter Marie-Laetitia,
with love, tenderness and affection

Series Editor
Roger Prud'homme

Stable Numerical Schemes for Fluids, Structures and their Interactions

Cornel Marius Murea

ISTE PRESS

ELSEVIER

First published 2017 in Great Britain and the United States by ISTE Press Ltd and Elsevier Ltd

ISTE Press Ltd
27-37 St George's Road
London SW19 4EU
UK

www.iste.co.uk

Elsevier Ltd
The Boulevard, Langford Lane
Kidlington, Oxford, OX5 1GB
UK

www.elsevier.com

British Library Cataloguing-in-Publication Data
A CIP record for this book is available from the British Library
Library of Congress Cataloging in Publication Data
A catalog record for this book is available from the Library of Congress
ISBN 978-1-78548-273-1

Printed and bound in the UK and US

Contents

Preface

This book presents stable numerical schemes for incompressible fluids, linear and nonlinear elastic structures, and fluid-structure interactions. It reviews a selection of fundamental methods as well as some new approaches.

For fluids governed by the Stokes equations, we present the mixed finite element method, and for the Navier-Stokes equations, three discretization schemes of order one in time will be analyzed. In cases where the domain occupied by the fluid is moving, the ALE (Arbitrary Lagrangian-Eulerian) formalism will be used, and we demonstrate the stability of conservative and non-conservative schemes. Unconditionally stable schemes are analyzed for linear and nonlinear elastic structures. For the fluid-structure interaction, we first examine an implicit scheme, and we then present a semi-implicit scheme where the fluid domain is calculated explicitly, but the velocity and the pressure of the fluid and the structure displacement are evaluated implicitly.

We give a detailed demonstration of stability for each scheme presented. Each chapter finishes with numerical tests that confirm the theoretical results.

The book is aimed at second-year Master's degree students in mathematics, as well as students at engineering schools. It is based on two courses that I have taught at the University of Strasbourg for second-year Master's degree students of mathematics.

I thank Professor Frédéric Hecht for FreeFem++, an extremely powerful and user-friendly software that I used to carry out the numerical tests.

I would like to express my gratitude to the production team at ISTE and to Professor Roger Prud'homme, the co-ordinator of the collection, for publishing this work and giving me the opportunity to fulfill a deeply held ambition.

I would like to express my thanks to the University of Upper Alsace for the six month "Leave for Research or Thematic Conversions", which allowed me to finish this project.

I thank my wife for her daily help and for giving me the time to write this book. I dedicate this work to my daughter for the constant happiness she brings me.

I thank my mother for her kindness and generosity, and my sister for her support and encouragement.

Cornel Marius MUREA
June 2017

1

Mixed Finite Elements for the Stokes Equations

1.1. Stokes equations

Let $\Omega \subset \mathbb{R}^2$ be an open, non-empty set. We denote its border with $\partial\Omega$. A general point in $\mathbb{R}^2$ is denoted by $\mathbf{x} = (x_1, x_2)$. The gradients of a scalar function $q : \Omega \to \mathbb{R}$, $q \in C^1(\Omega)$ and of a vector-valued function $\mathbf{w} = (w_1, w_2) : \Omega \to \mathbb{R}^2$, $\mathbf{w} \in \left(C^1(\Omega)\right)^2$ are denoted by

$$\nabla q = \begin{pmatrix} \dfrac{\partial q}{\partial x_1} \\ \dfrac{\partial q}{\partial x_2} \end{pmatrix}, \quad \nabla \mathbf{w} = \begin{pmatrix} \dfrac{\partial w_1}{\partial x_1} & \dfrac{\partial w_1}{\partial x_2} \\ \dfrac{\partial w_2}{\partial x_1} & \dfrac{\partial w_2}{\partial x_2} \end{pmatrix}.$$

The divergence operators of a vector-valued function $\mathbf{w} = (w_1, w_2) : \Omega \to \mathbb{R}^2$, $\mathbf{w} \in \left(C^1(\Omega)\right)^2$ and of a tensor $\boldsymbol{\sigma} = \left(\sigma_{ij}\right)_{1\le i,j\le 2}$, $\boldsymbol{\sigma} \in \left(C^1(\Omega)\right)^4$ are denoted by

$$\nabla \cdot \mathbf{w} = \frac{\partial w_1}{\partial x_1} + \frac{\partial w_2}{\partial x_2}, \quad \nabla \cdot \boldsymbol{\sigma} = \begin{pmatrix} \dfrac{\partial \sigma_{11}}{\partial x_1} + \dfrac{\partial \sigma_{12}}{\partial x_2} \\ \dfrac{\partial \sigma_{21}}{\partial x_1} + \dfrac{\partial \sigma_{22}}{\partial x_2} \end{pmatrix}.$$

The scalar product of two vectors $\mathbf{v}$, $\mathbf{w}$ in $\mathbb{R}^2$ is denoted by $\mathbf{v}\cdot\mathbf{w} = \sum_{i=1}^{2} v_i\, w_i$. If $\boldsymbol{\sigma} = \left(\sigma_{ij}\right)_{1\le i,j\le 2}$ and $\boldsymbol{\tau} = \left(\tau_{ij}\right)_{1\le i,j\le 2}$ are two tensors, we note

$$\boldsymbol{\sigma} : \boldsymbol{\tau} = \sum_{i=1}^{2}\sum_{j=1}^{2} \sigma_{ij}\tau_{ij}.$$

If $\mathbf{v} \in \left(C^1(\Omega)\right)^2$, we denote the (linearized) strain rate tensor by

$$\epsilon(\mathbf{v}) = \frac{1}{2}\left(\nabla\mathbf{v} + (\nabla\mathbf{v})^T\right).$$

We seek the velocity $\mathbf{v} = (v_1, v_2) : \overline{\Omega} \to \mathbb{R}^2$, $\mathbf{v} \in \left(C^2(\Omega)\right)^2 \cap \left(C^0(\overline{\Omega})\right)^2$ and the pressure $p : \overline{\Omega} \to \mathbb{R}$, $p \in C^1(\Omega) \cap C^0(\overline{\Omega})$, such that

$$-2\mu\nabla \cdot \epsilon(\mathbf{v}) + \nabla p = \mathbf{f}, \quad \text{in } \Omega, \qquad [1.1]$$

$$\nabla \cdot \mathbf{v} = 0, \quad \text{in } \Omega, \qquad [1.2]$$

$$\mathbf{v} = 0, \quad \text{on } \partial\Omega, \qquad [1.3]$$

where

- $\mu > 0$ is the dynamic viscosity,
- $\mathbf{f} = (f_1, f_2) : \Omega \to \mathbb{R}^2$ are the body forces, for example gravitational forces.

A solution $(\mathbf{v}, p)$ of [1.1]–[1.3] is called a *classical solution*. If $(\mathbf{v}, p)$ is a solution of [1.1]–[1.3], then $(\mathbf{v}, p + const)$ is also a solution. We say that the pressure is determined up to an additive constant. We can choose a representative pressure by setting $\int_\Omega p\, d\mathbf{x} = 0$.

Let $\mathbf{w} \in \left(C^1(\Omega)\right)^2 \cap \left(C^0(\overline{\Omega})\right)^2$ such that $\mathbf{w} = 0$ on $\partial\Omega$. By multiplying [1.1] by $\mathbf{w}$, integrating over Ω and applying Green's formula (see theorem A.5), we obtain

$$-2\mu \int_\Omega (\nabla \cdot \epsilon(\mathbf{v})) \cdot \mathbf{w}\, d\mathbf{x} + \int_\Omega \nabla p \cdot \mathbf{w}\, d\mathbf{x} = \int_\Omega \mathbf{f} \cdot \mathbf{w}\, d\mathbf{x}$$

$$\Leftrightarrow 2\mu \int_\Omega \epsilon(\mathbf{v}) : \nabla\mathbf{w}\, d\mathbf{x} - \int_\Omega (\nabla \cdot \mathbf{w})\, p\, d\mathbf{x} = \int_\Omega \mathbf{f} \cdot \mathbf{w}\, d\mathbf{x}.$$

If A, B are two tensors and A is symmetric, then by using the identity $A : B = A^T : B^T$,

$$A : B = \frac{1}{2}A : B + \frac{1}{2}A^T : B^T = \frac{1}{2}A : B + \frac{1}{2}A : B^T = A : \frac{1}{2}\left(B + B^T\right).$$

As a result, we obtain $\boldsymbol{\epsilon}(\mathbf{v}) : \nabla\mathbf{w} = \boldsymbol{\epsilon}(\mathbf{v}) : \boldsymbol{\epsilon}(\mathbf{w})$. We thus have

$$2\mu \int_\Omega \boldsymbol{\epsilon}(\mathbf{v}) : \boldsymbol{\epsilon}(\mathbf{w})\, d\mathbf{x} - \int_\Omega (\nabla \cdot \mathbf{w})\, p\, d\mathbf{x} = \int_\Omega \mathbf{f} \cdot \mathbf{w}\, d\mathbf{x}$$

for all $\mathbf{w} \in \left(C^1(\Omega)\right)^2 \cap \left(C^0(\overline{\Omega})\right)^2$ such that $\mathbf{w} = 0$ on $\partial\Omega$.

By multiplying [1.2] by $q \in C^0(\overline{\Omega})$ satisfying $\int_\Omega q\, d\mathbf{x} = 0$ and integrating over Ω, we obtain

$$\int_\Omega (\nabla \cdot \mathbf{v})\, q\, d\mathbf{x} = 0.$$

Evidently, we have not used $\int_\Omega q\, d\mathbf{x} = 0$ to obtain the equation above.

In order to demonstrate the existence and uniqueness of a solution of the Stokes equations, we will use Sobolev spaces (see Appendix A.1). The scalar products and norms of the spaces $L^2(\Omega)$, $H^1(\Omega)$ are denoted by

$$(v, w)_{0,\Omega} = \int_\Omega v(\mathbf{x})w(\mathbf{x})\, d\mathbf{x}, \quad ||v||_{0,\Omega} = \sqrt{(v, v)_{0,\Omega}}$$

and

$$(v, w)_{1,\Omega} = \int_\Omega v(\mathbf{x})w(\mathbf{x})\, d\mathbf{x} + \int_\Omega \nabla v \cdot \nabla w\, d\mathbf{x}, \quad ||v||_{1,\Omega} = \sqrt{(v, v)_{1,\Omega}}$$

respectively.

We assume that $\Omega \subset \mathbb{R}^2$ is an open, non-empty, connected, bounded set and its boundary $\partial\Omega$ is Lipschitz (see definition A.1 in the Appendix). We note

$$L_0^2(\Omega) = \{q \in L^2(\Omega);\ \int_\Omega q\, d\mathbf{x} = 0\}.$$

To ease the notation, we will set $W = (H_0^1(\Omega))^2$ and $Q = L_0^2(\Omega)$. We introduce the space of divergence-free functions

$$V = \{\mathbf{w} \in (H_0^1(\Omega))^2;\ \nabla \cdot \mathbf{w} = 0\}.$$

We assume that $\mathbf{f} \in (L^2(\Omega))^2$. Find $\mathbf{v} \in W$ and $p \in Q$ such that

$$2\mu \int_\Omega \epsilon(\mathbf{v}) : \epsilon(\mathbf{w})\, d\mathbf{x} - \int_\Omega (\nabla \cdot \mathbf{w})\, p\, d\mathbf{x} = \int_\Omega \mathbf{f} \cdot \mathbf{w}\, d\mathbf{x}, \forall \mathbf{w} \in W \qquad [1.4]$$

$$-\int_\Omega (\nabla \cdot \mathbf{v})\, q\, d\mathbf{x} = 0, \ \forall q \in Q. \qquad [1.5]$$

System [1.4]–[1.5] is called *the mixed weak formulation* of the Stokes equations.

We can introduce a second weak formulation using the space of divergence-free functions: find $\mathbf{v} \in V$ such that

$$2\mu \int_\Omega \epsilon(\mathbf{v}) : \epsilon(\mathbf{w})\, d\mathbf{x} = \int_\Omega \mathbf{f} \cdot \mathbf{w}\, d\mathbf{x}, \quad \forall \mathbf{w} \in V. \qquad [1.6]$$

If $(\mathbf{v}, p)$ is a solution of [1.4]–[1.5], then $\mathbf{v}$ is a solution of [1.6]. The mixed weak formulation is preferable for finite element approximations. Instead of using test divergence-free functions $\mathbf{w} \in V$ which are difficult to approximate, the mixed weak formulation uses the test functions $\mathbf{w} \in W$.

1.2. Existence and uniqueness for an abstract mixed problem

Let W, Q be two Hilbert spaces with scalar products $(\cdot,\cdot)_W$, $(\cdot,\cdot)_Q$ and norms $\|\cdot\|_W$, $\|\cdot\|_Q$, respectively. Let $b : W \times Q \to \mathbb{R}$ be a bilinear function. We assume that b is continuous and we have

$$\exists M_b > 0, \ \forall w \in W, \ \forall q \in Q, \quad |b(w,q)| \leq M_b \|w\|_W \|q\|_Q. \qquad [1.7]$$

We use Q' to denote the dual space of Q, *i.e.* the space of linear, continuous functions on Q with values in $\mathbb{R}$. There exists the linear, continuous operator $B : W \to Q'$, such that

$$\forall w \in W, \forall q \in Q, \quad b(w,q) = \langle Bw, q \rangle_{Q',Q}$$

where $\langle \cdot,\cdot \rangle_{Q',Q}$ is the duality product of Q' and Q. We can define the linear, continuous adjoint operator $B^* : Q \to W'$, such that

$$\forall w \in W, \forall q \in Q, \quad \langle Bw, q \rangle_{Q',Q} = \langle w, B^* q \rangle_{W,W'}.$$

We note

$$V = \{w \in W;\ \forall q \in Q,\ b(w,q) = 0\}$$

$$V^{\perp} = \{w \in W;\ \forall v \in V,\ (w,v)_W = 0\}$$

$$V^{\circ} = \{f \in W';\ \forall v \in V,\ \langle f, v\rangle_{W',W} = 0\}.$$

Since b is continuous, V is closed in W and we have the decomposition $W = V \oplus V^{\perp}$, *i.e.* for all $w \in W$, there exist unique $w^0 \in V$ and $w^{\perp} \in V^{\perp}$ such that $w = w^0 + w^{\perp}$ (see [DAU 88], vol. 4, Chap. VI, Proposition 11 and Remark 14, p. 1122–1124). Using the Riesz-Fréchet representation theorem A.8 in the Appendix, we can construct a bijection between V° and $(V^{\perp})'$.

LEMMA 1.1.– *Let $b : W \times Q \to \mathbb{R}$ be a bilinear, continuous function. Let the following statements:*

i) there exists a constant $\beta > 0$ such that

$$\inf_{q\in Q, q\neq 0}\ \sup_{w\in W, w\neq 0} \frac{b(w,q)}{\|w\|_W \|q\|_Q} \geq \beta$$

ii) the operator $B^* : Q \to V^{\circ} \subset W'$ is bijective and $\|B^* q\|_{W'} \geq \beta \|q\|_Q$ for all $q \in Q$

iii) the operator $B : V^{\perp} \to Q'$ is bijective and $\left\|Bv^{\perp}\right\|_{Q'} \geq \beta \left\|v^{\perp}\right\|_W$ for all $v^{\perp} \in V^{\perp}$.

Then i) $\Rightarrow$ ii) $\Rightarrow$ iii).

DEMONSTRATION 1.1.– i) $\Rightarrow$ ii). Let $q \in Q$, $q \neq 0$. We have

$$\frac{1}{\|q\|_Q} \sup_{w\in W, w\neq 0} \frac{b(w,q)}{\|w\|_W} = \sup_{w\in W, w\neq 0} \frac{b(w,q)}{\|w\|_W \|q\|_Q} \geq \inf_{q\in Q, q\neq 0}\ \sup_{w\in W, w\neq 0} \frac{b(w,q)}{\|w\|_W \|q\|_Q} \geq \beta.$$

By construction, $b(w,q) = \langle w, B^* q\rangle_{W,W'}$ and by definition

$$\left\|B^* q\right\|_{W'} \overset{déf}{=} \sup_{w\in W, w\neq 0} \frac{\langle w, B^* q\rangle_{W,W'}}{\|w\|_W}.$$

Thus,

$$\frac{1}{\|q\|_Q}\left\|B^*q\right\|_{W'} \geq \beta \Rightarrow \left\|B^*q\right\|_{W'} \geq \beta \|q\|_Q, \quad \forall q \in Q,\ q \neq 0.$$

Evidently, the inequality above is also valid for $q = 0$. We will show that $B^* : Q \to W'$ is injective by assuming the opposite. If $q_1, q_2 \in Q$ such that $B^*q_1 = B^*q_2$, then $0 = B^*(q_1 - q_2)$ and $0 = \|B^*(q_1 - q_2)\|_{W'} \geq \beta \|(q_1 - q_2)\|_Q \geq 0$. Thus, $q_1 = q_2$ and B^* is injective.

Since B is continuous, it is a closed operator and we can apply point iv) of theorem A.6 in the Appendix. We obtain $R(B^*) = (N(B))^\circ = V^\circ$. Thus, B^* is surjective from Q onto V°, and consequently, $B^* : Q \to V^\circ$ is bijective.

ii) $\Rightarrow$ iii). Since B is closed and $\|B^*q\|_{W'} \geq \beta \|q\|_Q$, for all $q \in Q$, by applying theorem A.7 in the Appendix, we find that $B : W \to Q'$ is surjective. Let $g \in Q'$; thus, there exists $v \in W$ such that $Bv = g$. Since $W = V \oplus V^\perp$, $v = v^0 + v^\perp$ with $v^0 \in V$ and $v^\perp \in V^\perp$. Hence, $g = Bv = B(v^0 + v^\perp) = Bv^\perp$ because $Bv^0 = 0$. Thus, the operator $B : V^\perp \to Q'$ is surjective.

Let $v_1^\perp, v_2^\perp \in V^\perp$ such that $Bv_1^\perp = Bv_2^\perp$. Hence, $B(v_1^\perp - v_2^\perp) = 0$, and thus, $v_1^\perp - v_2^\perp \in V$. However, $V^\perp$ is a vector subspace, thus $v_1^\perp - v_2^\perp \in V^\perp$. Since $V^\perp \cap V = \{0\}$, $v_1^\perp = v_2^\perp$; thus, $B : V^\perp \to Q'$ is injective.

Let $v^\perp \in V^\perp$, $v^\perp \neq 0$. We use the same principle as the demonstration of Prop. 4.1.5, p. 213, [BOF 13]. We define $f \in W'$ by $\langle f, w\rangle_{W',W} = (v^\perp, w)_W$ which satisfies $\left\|v^\perp\right\|_W = \|f\|_{W'}$. Furthermore, $f \in V^\circ$ and $f \neq 0$. Using ii), there exists a unique $q \in Q$ such that $B^*q = f$ and $q \neq 0$. From the construction of f, we have

$$\|f\|_{W'} \times \left\|v^\perp\right\|_W = \left\|v^\perp\right\|_W^2 = (v^\perp, v^\perp)_W = \langle f, v^\perp\rangle_{W',W}.$$

By the definition of an adjoint operator, we have

$$\langle f, v^\perp\rangle_{W',W} = \langle B^*q, v^\perp\rangle_{W',W} = \langle q, Bv^\perp\rangle_{Q,Q'}$$

and by using the Cauchy-Schwarz inequality, we obtain

$$\langle q, Bv^\perp\rangle_{Q,Q'} \leq \|q\|_Q \left\|Bv^\perp\right\|_{Q'}.$$

We can use the inequality of point ii) $\|B^*q\|_{W'} \geq \beta \|q\|_Q$ for all $q \in Q$ and we obtain

$$\beta \|q\|_Q \left\|v^\perp\right\|_W \leq \left\|B^* q\right\|_{W'} \times \left\|v^\perp\right\|_W = \|f\|_{W'} \times \left\|v^\perp\right\|_W \leq \|q\|_Q \left\|Bv^\perp\right\|_{Q'}$$

from which $\beta \left\|v^\perp\right\|_W \leq \left\|Bv^\perp\right\|_{Q'}$ for all $v^\perp \in V^\perp$, $v^\perp \neq 0$. For $v^\perp = 0$, the inequality is evident. □

REMARK 1.1.– *We can show that i), ii) and iii) are equivalent (see [GIR 86], Lemme 4.1, p. 58.)*

THEOREM 1.1.– *[BRE 74] Let $a : W \times W \to \mathbb{R}$ be a bilinear, continuous map. We assume that a is V-elliptic,* i.e.

$$\exists \alpha > 0, \ \forall v \in V, \quad \alpha \|v\|_W^2 \leq a(v, v). \qquad [1.8]$$

Let $b : W \times Q \to \mathbb{R}$ be a bilinear, continuous map such that

$$\exists \beta > 0, \qquad \inf_{q \in Q, q \neq 0} \ \sup_{w \in W, w \neq 0} \frac{b(w, q)}{\|w\|_W \|q\|_Q} \geq \beta. \qquad [1.9]$$

Let $f \in W'$ and $S \in Q'$, then the problem: find $v \in W$ and $p \in Q$

$$a(v, w) + b(w, p) = \langle f, w \rangle_{W',W}, \quad \forall w \in W \qquad [1.10]$$

$$b(v, q) = \langle S, q \rangle_{Q',Q}, \quad \forall q \in Q \qquad [1.11]$$

has a unique solution.

DEMONSTRATION 1.2.– We present an adaptation of the proof given by J. M. Thomas in [THO 77]. Since $W = V \oplus V^\perp$, for all $w \in W$, there exist unique $w^0 \in V$ and $w^\perp \in V^\perp$ such that $w = w^0 + w^\perp$. Similarly, there exist unique $v^0 \in V$ and $v^\perp \in V^\perp$ such that $v = v^0 + v^\perp$. The system [1.10]–[1.11] is equivalent to

$$a(v^0 + v^\perp, w^0 + w^\perp) + b(w^0 + w^\perp, p)$$
$$= \langle f, w^0 + w^\perp \rangle_{W',W}, \forall w^0 \in V, \forall w^\perp \in V^\perp$$
$$b(v^0 + v^\perp, q) = \langle S, q \rangle_{Q',Q}, \ \forall q \in Q.$$

Since $b(w^0, p) = 0$ and $b(v^0, q) = 0$, we then obtain the equivalent system

$$a(v^0, w^\perp) + a(v^\perp, w^\perp) + b(w^\perp, p) = \langle f, w^\perp \rangle_{W',W}, \ \forall w^\perp \in V^\perp \quad [1.12]$$

$$a(v^0, w^0) + a(v^\perp, w^0) = \langle f, w^0 \rangle_{W',W}, \ \forall w^0 \in V \quad [1.13]$$

$$b(v^\perp, q) = \langle S, q \rangle_{Q',Q}, \ \forall q \in Q. \quad [1.14]$$

Evidently, by summing [1.12] and [1.13] and taking account of $b(w^0, p) = 0$, we obtain [1.10]. The system [1.12], [1.13], [1.14] is equivalent to the system [1.10], [1.13], [1.14]. This latter system will be used to find the unique solution of the initial system [1.10]–[1.11].

Equation [1.14] can be equivalently expressed as

$$\langle Bv^\perp, q \rangle_{Q',Q} = \langle S, q \rangle_{Q',Q}, \ \forall q \in Q \Leftrightarrow Bv^\perp = S \text{ in } Q'.$$

Using lemma 1.1, point iii), there exists a unique $v^\perp \in V^\perp$ such that $Bv^\perp = S$. Since $v^\perp$ is known, we can find unique $v^0 \in V$ from [1.13], using the Lax-Milgram theorem A.9 in the Appendix. Essentially, [1.13] can be written as

$$a(v^0, w^0) = \langle f, w^0 \rangle_{W',W} - a(v^\perp, w^0), \ \forall w^0 \in V$$

and a is bilinear, continuous and V-elliptic. Finally, we will find unique $p \in Q$, starting from [1.10], which is equivalent to

$$b(w, p) = \langle f, w \rangle_{W',W} - a(v, w), \ \forall w \in W.$$

Let $\ell : W \to \mathbb{R}$, defined by $\ell(w) = \langle f, w \rangle_{W',W} - a(v, w)$. We have $\ell \in W'$. Taking account of [1.13], we obtain $\ell \in V^\circ$.

Using the adjoint operator B^*, we obtain the equivalent equation to [1.10]: find $p \in Q$ such that

$$\langle w, B^* p \rangle_{W,W'} = \ell(w), \ \forall w \in W \Leftrightarrow B^* p = \ell \text{ in } W'.$$

Using Lemma 1.1, point ii), the operator $B^* : Q \to V^\circ$ is bijective; therefore, for every $\ell \in V^\circ$, there exists a unique $p \in Q$ such that $B^* p = \ell$ in W'. We have demonstrated the existence and uniqueness of $v = v^0 + v^\perp \in W$ and $p \in Q$, solution of [1.10]–[1.11]. □

REMARK 1.2.–

i) The map a is not necessarily symmetric.

ii) Relation [1.9] is called the inf-sup condition, *or the Babuska-Brezzi condition, or the Ladyjenskaia-Babuska-Brezzi condition. Later, we will introduce the discrete inf-sup condition and to avoid confusion, we will call relation [1.9] the* continuous inf-sup condition.

PROPOSITION 1.1.– *[continuous dependence on parameters] In the hypotheses of theorem 1.1, we have that the solution $v \in W$ and $p \in Q$ satisfies:*

$$\|v\|_W \leq \frac{1}{\alpha}\|f\|_{W'} + \frac{1}{\beta}\left(1+\frac{M_a}{\alpha}\right)\|S\|_{Q'} \qquad [1.15]$$

$$\|p\|_Q \leq \frac{1}{\beta}\left(1+\frac{M_a}{\alpha}\right)\|f\|_{W'} + \frac{M_a}{\beta^2}\left(1+\frac{M_a}{\alpha}\right)\|S\|_{Q'} \qquad [1.16]$$

where M_a is the continuity constant of a, or more precisely,

$$\exists M_a > 0,\ \forall w_1 \in W,\ \forall w_2 \in W, \quad |a(w_1, w_2)| \leq M_a \|w_1\|_W \|w_2\|_W . \quad [1.17]$$

DEMONSTRATION 1.3.– Using the notation of the demonstration of theorem 1.1, we have

$$\begin{aligned} Bv^{\perp} &= S \text{ in } Q' \\ a(v^0, w^0) &= \langle f, w^0\rangle_{W',W} - a(v^{\perp}, w^0),\ \forall w^0 \in V \\ B^* p &= \ell \text{ in } W' \end{aligned}$$

where $\ell \in W'$, $\ell(w) = \langle f, w\rangle_{W',W} - a(v, w)$. Using lemma 1.1, point iii), we obtain

$$\beta \left\|v^{\perp}\right\|_W \leq \left\|Bv^{\perp}\right\|_{Q'} = \|S\|_{Q'} \Leftrightarrow \left\|v^{\perp}\right\|_W \leq \frac{1}{\beta}\|S\|_{Q'} .$$

The continuity of a gives us

$$\left|\langle f, w^0\rangle_{W',W} - a(v^{\perp}, w^0)\right| \leq \|f\|_{W'} \left\|w^0\right\|_W + M_a \left\|v^{\perp}\right\|_W \left\|w^0\right\|_W$$

and using the Lax-Milgram theorem A.9 in the Appendix, we have

$$\left\|v^0\right\|_W \leq \frac{1}{\alpha}\left(\|f\|_{W'} + M_a \left\|v^\perp\right\|_W\right).$$

Thus,

$$\|v\|_W \leq \left\|v^\perp\right\|_W + \left\|v^0\right\|_W \leq \frac{1}{\beta}\|S\|_{Q'} + \frac{1}{\alpha}\left(\|f\|_{W'} + M_a\frac{1}{\beta}\|S\|_{Q'}\right)$$

which gives us [1.15].

We have that $\|\ell\|_{W'} \leq \|f\|_{W'} + M_a \|v\|_W$ and using lemma 1.1, point ii), we have

$$\begin{aligned}
\beta \|p\|_Q &\leq \left\|B^* p\right\|_{W'} = \|\ell\|_{W'} \leq \|f\|_{W'} + M_a \|v\|_W \\
&\leq \|f\|_{W'} + \frac{M_a}{\alpha}\|f\|_{W'} + \frac{M_a}{\beta}\left(1 + \frac{M_a}{\alpha}\right)\|S\|_{Q'} \\
&= \left(1 + \frac{M_a}{\alpha}\right)\|f\|_{W'} + \frac{M_a}{\beta}\left(1 + \frac{M_a}{\alpha}\right)\|S\|_{Q'}
\end{aligned}$$

which gives us [1.16]. □

REMARK 1.3.– *The map a is bilinear and continuous, so there exists a linear, continuous operator $A : W \to W'$ such that $a(w_1, w_2) = \langle Aw_1, w_2\rangle_{W',W}$, for all $w_1, w_2 \in W$. System [1.10]–[1.11] can be written in the equivalent form*

$$Av + B^* p = f \text{ in } W' \qquad [1.18]$$

$$Bv = S \text{ in } Q'. \qquad [1.19]$$

Proposition 1.1 means that if α and β are not close to zero, then if we make small errors in the values of f and S, we will obtain small errors in the solution v and p.

1.3. Existence and uniqueness for the Stokes equations

We assume that $\Omega \subset \mathbb{R}^2$ is an open, non-empty, bounded, connected set and its boundary $\partial\Omega$ is Lipschitz. We note $W = (H_0^1(\Omega))^2$, $Q = L_0^2(\Omega)$ and

$$V = \{\mathbf{w} \in (H_0^1(\Omega))^2;\ \nabla \cdot \mathbf{w} = 0\}.$$

We have seen that we can write the Stokes equations in the form of a mixed problem [1.4]–[1.5]. We note

$$a : W \times W \to \mathbb{R}, \quad a(\mathbf{v}, \mathbf{w}) = \int_\Omega 2\mu\, \boldsymbol{\epsilon}(\mathbf{v}) : \boldsymbol{\epsilon}(\mathbf{w})\, d\mathbf{x},$$

$$b : W \times Q \to \mathbb{R}, \quad b(\mathbf{v}, q) = -\int_\Omega (\nabla \cdot \mathbf{v})\, q\, d\mathbf{x}.$$

THEOREM 1.2.–

i) The map a is bilinear, continuous and W-elliptic, i.e.

$$\exists \alpha > 0,\ \forall w \in W, \quad \alpha \|w\|_W^2 \le a(w, w). \qquad [1.20]$$

ii) The map b is bilinear and continuous, such that

$$\exists \beta > 0, \quad \inf_{q \in Q, q \neq 0} \sup_{w \in W, w \neq 0} \frac{b(w, q)}{\|w\|_W \|q\|_Q} \ge \beta. \qquad [1.21]$$

iii) Let $f \in W'$. Hence, the problem: find $v \in W$ and $p \in Q$

$$a(v, w) + b(w, p) = \langle f, w \rangle_{W', W}, \quad \forall w \in W \qquad [1.22]$$

$$b(v, q) = 0, \quad \forall q \in Q \qquad [1.23]$$

has a unique solution.

DEMONSTRATION 1.4.–

i) Since $\mathbf{w} \to \boldsymbol{\epsilon}(\mathbf{w})$ is linear, a is bilinear. We will show the continuity. We have $\boldsymbol{\epsilon}(\mathbf{v}) = (\epsilon_{ij}(\mathbf{v}))_{1 \le i,j \le 2}$ where $\epsilon_{ij}(\mathbf{v}) = \frac{1}{2}\left(\frac{\partial v_i}{\partial x_j} + \frac{\partial v_j}{\partial x_i}\right)$. By applying the Cauchy-Schwarz inequality [A.1], we have

$$|a(\mathbf{v}, \mathbf{w})| = \left| 2\mu \sum_{i,j=1}^{2} \int_\Omega \epsilon_{ij}(\mathbf{v})\epsilon_{ij}(\mathbf{w}) d\mathbf{x} \right| \le 2\mu \sum_{i,j=1}^{2} \left\|\epsilon_{ij}(\mathbf{v})\right\|_{0,\Omega} \left\|\epsilon_{ij}(\mathbf{w})\right\|_{0,\Omega}.$$

Using [A.2], we obtain

$$\sum_{i,j=1}^{2} \left\|\epsilon_{ij}(\mathbf{v})\right\|_{0,\Omega} \left\|\epsilon_{ij}(\mathbf{w})\right\|_{0,\Omega} \le \sqrt{\sum_{i,j=1}^{2} \left\|\epsilon_{ij}(\mathbf{v})\right\|_{0,\Omega}^2} \sqrt{\sum_{i,j=1}^{2} \left\|\epsilon_{ij}(\mathbf{w})\right\|_{0,\Omega}^2}.$$

We have

$$\left\|\epsilon_{ij}(\mathbf{v})\right\|_{0,\Omega} = \frac{1}{2}\left\|\frac{\partial v_i}{\partial x_j} + \frac{\partial v_j}{\partial x_i}\right\|_{0,\Omega} \leq \frac{1}{2}\left(\left\|\frac{\partial v_i}{\partial x_j}\right\|_{0,\Omega} + \left\|\frac{\partial v_j}{\partial x_i}\right\|_{0,\Omega}\right)$$

and by using $(\frac{r+s}{2})^2 \leq \frac{r^2+s^2}{2}$, if $r, s \in \mathbb{R}$, we obtain

$$\left\|\epsilon_{ij}(\mathbf{v})\right\|_{0,\Omega}^2 \leq \frac{1}{2}\left(\left\|\frac{\partial v_i}{\partial x_j}\right\|_{0,\Omega}^2 + \left\|\frac{\partial v_j}{\partial x_i}\right\|_{0,\Omega}^2\right)$$

from which

$$\sum_{i,j=1}^{2}\left\|\epsilon_{ij}(\mathbf{v})\right\|_{0,\Omega}^2 \leq \frac{1}{2}\left(\sum_{i,j=1}^{2}\left\|\frac{\partial v_i}{\partial x_j}\right\|_{0,\Omega}^2 + \sum_{i,j=1}^{2}\left\|\frac{\partial v_j}{\partial x_i}\right\|_{0,\Omega}^2\right) = \|\nabla \mathbf{v}\|_{0,\Omega}^2 .$$

Finally, we have

$$|a(\mathbf{v}, \mathbf{w})| \leq 2\mu \, \|\nabla \mathbf{v}\|_{0,\Omega} \, \|\nabla \mathbf{w}\|_{0,\Omega} \leq 2\mu \, \|\mathbf{v}\|_{1,\Omega} \, \|\mathbf{w}\|_{1,\Omega}$$

which proves the continuity of a.

The W-ellipticity result of Korn's theorem A.4

$$a(\mathbf{w}, \mathbf{w}) = 2\mu \sum_{i,j=1}^{2} \int_\Omega \left(\epsilon_{ij}(\mathbf{w})\right)^2 d\mathbf{x} = 2\mu \, \|\epsilon\,(\mathbf{w})\|_{0,\Omega}^2 \geq \frac{2\mu}{C_K(\Omega)^2} \, \|\mathbf{w}\|_{1,\Omega}^2 .$$

ii) Since $\mathbf{w} \to \nabla \cdot \mathbf{w}$ is linear, b is bilinear. We will show the continuity. Using the Cauchy-Schwarz inequality, we have

$$|b(\mathbf{w}, q)| = \left|\sum_{i=1}^{2} \int_\Omega \frac{\partial w_i}{\partial x_i} q \, d\mathbf{x}\right| \leq \sum_{i=1}^{2} \left|\int_\Omega \frac{\partial w_i}{\partial x_i} q \, d\mathbf{x}\right| \leq \sum_{i=1}^{2} \left\|\frac{\partial w_i}{\partial x_i}\right\|_{0,\Omega} \|q\|_{0,\Omega}$$

$$= \left(\sum_{i=1}^{2} \left\|\frac{\partial w_i}{\partial x_i}\right\|_{0,\Omega}\right) \|q\|_{0,\Omega} .$$

Since $(\frac{r+s}{2})^2 \leq \frac{r^2+s^2}{2}$, then

$$\sum_{i=1}^{2} \left\| \frac{\partial w_i}{\partial x_i} \right\|_{0,\Omega} \leq \sqrt{2} \sqrt{\sum_{i=1}^{2} \left\| \frac{\partial w_i}{\partial x_i} \right\|_{0,\Omega}^2} \leq \sqrt{2}\, ||\nabla \mathbf{w}||_{0,\Omega}\,.$$

Finally, we have

$$|b(\mathbf{w}, q)| \leq \sqrt{2}\, ||\nabla \mathbf{w}||_{0,\Omega}\, ||q||_{0,\Omega} \leq \sqrt{2}\, ||\mathbf{w}||_{1,\Omega}\, ||q||_{0,\Omega}$$

which proves the continuity of b. This leaves us to show the inf-sup condition [1.21]. The operator $B : W \to Q'$ defined by $\langle B\mathbf{w}, q\rangle_{Q',Q} = (\nabla\cdot\mathbf{w}, q)_Q = b(\mathbf{w}, q)$ is the composition of the operator $\mathbf{w} \to \nabla \cdot \mathbf{w}$ from W to Q and the canonical injection j from Q to Q' (see definition A.5); thus, B is linear and continuous. Thanks to Corollary 2.4, p. 24 [GIR 86], the operator $\mathbf{w} \to \nabla \cdot \mathbf{w}$ from W to Q is surjective. By using the Riesz-Fréchet Representation theorem (see A.8), the canonical injection of Q into Q' is bijective. We obtain that B is surjective. We can define the adjoint operator $B^* : Q \to W'$, by

$$\langle B\mathbf{w}, q\rangle_{Q',Q} = \langle \mathbf{w}, B^* q\rangle_{W,W'}$$

and using the characterization theorem of surjective operators A.7, there exists $\beta > 0$ such that

$$\left\|B^* q\right\|_{W'} \geq \beta\, ||q||_Q\,.$$

However,

$$\left\|B^* q\right\|_{W'} \overset{déf}{=} \sup_{\mathbf{w}\in W, \mathbf{w}\neq 0} \frac{\langle \mathbf{w}, B^* q\rangle_{W,W'}}{||\mathbf{w}||_W} = \sup_{\mathbf{w}\in W, \mathbf{w}\neq 0} \frac{b(\mathbf{w}, q)}{||\mathbf{w}||_W}$$

thus,

$$\forall q \in Q, q \neq 0, \qquad \sup_{\mathbf{w}\in W, \mathbf{w}\neq 0} \frac{b(\mathbf{w}, q)}{||\mathbf{w}||_W\, ||q||_Q} \geq \beta$$

which implies

$$\inf_{q\in Q, q\neq 0} \sup_{w\in W, w\neq 0} \frac{b\,(w, q)}{||w||_W\, ||q||_Q} \geq \beta.$$

iii) We apply theorem 1.1 with $S = 0$. □

REMARK 1.4.– *A sufficient condition to have the inf-sup condition is that the operator* $\mathbf{w} \to \nabla \cdot \mathbf{w}$ *from* $(H_0^1(\Omega))^2$ *to* $L_0^2(\Omega)$ *is surjective.*

1.4. Finite element approximation of an abstract mixed problem

Let W, Q be two Hilbert spaces with scalar products $(\cdot,\cdot)_W$, $(\cdot,\cdot)_Q$ and norms $\|\cdot\|_W$, $\|\cdot\|_Q$, respectively. Let W_h, Q_h be two finite-dimension vector spaces such that $W_h \subset W$ and $Q_h \subset Q$. The scalar product $(\cdot,\cdot)_W$ induces on W_h a Hilbert space structure. Similarly, Q_h is a Hilbert space with scalar product $(\cdot,\cdot)_Q$. To approximate the solution of the mixed problem [1.10]–[1.11], we will introduce the following *discrete mixed problem*: find $v_h \in W_h$ and $p_h \in Q_h$ such that

$$a(v_h, w_h) + b(w_h, p_h) = \langle f, w_h \rangle_{W',W}, \quad \forall w_h \in W_h \qquad [1.24]$$

$$b(v_h, q_h) = \langle S, q_h \rangle_{Q',Q}, \quad \forall q_h \in Q_h \qquad [1.25]$$

where $f \in W'$ and $S \in Q'$. We recall that, for the existence and uniqueness of [1.10]–[1.11], which we will call the *continuous mixed problem*, we assumed that $a : W \times W \to \mathbb{R}$ is a bilinear, continuous map, more precisely,

$$\exists M_a > 0,\ \forall w_1 \in W,\ \forall w_2 \in W, \quad |a(w_1, w_2)| \leq M_a \|w_1\|_W \|w_2\|_W, \qquad [1.26]$$

V-elliptic (see [1.20]) and that $b : W \times Q \to \mathbb{R}$ is a bilinear, continuous map, more precisely,

$$\exists M_b > 0,\ \forall w \in W,\ \forall q \in Q, \quad |b(w, q)| \leq M_b \|w\|_W \|q\|_Q, \qquad [1.27]$$

which satisfies the continuous inf-sup condition, *i.e.* relation [1.9]. We denote the dual of W_h with W_h' and the dual of Q_h with Q_h', and

$$V_h = \{w_h \in W_h;\ \forall q_h \in Q_h,\ b(w_h, q_h) = 0\}$$

$$V_h(S) = \{w_h \in W_h;\ \forall q_h \in Q_h,\ b(w_h, q_h) = \langle S, q_h \rangle_{Q',Q}\}$$

$$V_h^{\perp} = \{w_h \in W_h;\ \forall v_h \in V_h,\ (w_h, v_h)_W = 0\}$$

$$V_h^{\circ} = \{f_h \in W_h';\ \forall v_h \in V_h,\ \langle f_h, v_h \rangle_{W',W} = 0\}.$$

We draw attention to the fact that V_h is not a subspace of V, thus we cannot replace V by V_h in [1.20], which means that we cannot obtain V_h-ellipticity from the V-ellipticity. The continuous inf-sup condition is equivalent to

$$\exists\beta > 0, \forall q \in Q, \qquad \sup_{w \in W, w \neq 0} \frac{b(w,q)}{\|w\|_W} \geq \beta \|q\|_Q .$$

Since $\sup_{w_h \in W_h, w_h \neq 0} \frac{b(w_h,q)}{\|w_h\|_W}$ is smaller than $\sup_{w \in W, w \neq 0} \frac{b(w,q)}{\|w\|_W}$, we cannot replace W by W_h in the above inequality.

If $w_h \in W_h$, then $\|w_h\|_W = \|w_h\|_{W_h}$ whereas, if $f \in W'$, then $f \in W_h'$, but

$$\|f\|_W \stackrel{def}{=} \sup_{w \in W, w \neq 0} \frac{\left|\langle f, w\rangle_{W',W}\right|}{\|w\|_W} \geq \|f\|_{W_h} \stackrel{def}{=} \sup_{w_h \in W_h, w_h \neq 0} \frac{\left|\langle f, w_h\rangle_{W_h',W_h}\right|}{\|w_h\|_{W_h}}.$$

THEOREM 1.3.– *Let $a : W \times W \to \mathbb{R}$ be a bilinear, continuous map (see [1.26]) and V-elliptic (see [1.8]). Furthermore, we assume that*

$$\exists\alpha^* > 0, \ \forall v_h \in V_h, \quad \alpha^* \|v_h\|_W^2 \leq a(v_h, v_h). \qquad [1.28]$$

Let $b : W \times Q \to \mathbb{R}$ be a bilinear, continuous map (see [1.27]), which satisfies [1.9]. Furthermore, we assume that b satisfies the discrete inf-sup *condition,* i.e.

$$\exists\beta^* > 0, \qquad \inf_{q_h \in Q, q_h \neq 0} \sup_{w_h \in W_h, w_h \neq 0} \frac{b(w_h, q_h)}{\|w_h\|_W \|q_h\|_Q} \geq \beta^*. \qquad [1.29]$$

Let $f \in W'$ and $S \in Q'$, then problem [1.24]–[1.25] has a unique solution and there exists a constant C dependent on α^*, β^*, M_a, M_b such that

$$\|v - v_h\|_W + \|p - p_h\|_Q \leq C\left(\inf_{w_h \in W_h} \|v - w_h\|_W + \inf_{q_h \in Q_h} \|p - q_h\|_W \right). \qquad [1.30]$$

DEMONSTRATION 1.5.– We deduce from [1.26] that

$$\forall w_h \in W_h, \ \forall u_h \in W_h, \quad |a(w_h, u_h)| \leq M_a \|w_h\|_W \|u_h\|_W$$

thus the map $a : W_h \times W_h \to \mathbb{R}$ is continuous. Similarly, $b : W_h \times Q_h \to \mathbb{R}$ is continuous. The map $w_h \in W_h \to \langle f, w_h\rangle_{W',W}$ is an element $f_h \in W_h'$ and the map $q_h \in Q_h \to \langle S, q_h \rangle_{Q',Q}$ is an element $S_h \in Q'_h$. The existence and uniqueness of problem [1.24]–[1.25] is a consequence of theorem 1.1 for the Hilbert spaces W_h, Q_h, the maps $a : W_h \times W_h \to \mathbb{R}$, $b : W_h \times Q_h \to \mathbb{R}$ and the values $f_h \in W'_h$, $S_h \in Q'_h$.

We will show [1.30] in three stages:

$$i) \inf_{\overline{w}_h \in V_h(S)} \|v - \overline{w}_h\|_W \le \left(1 + \frac{M_b}{\beta^*}\right) \inf_{w_h \in W_h} \|v - w_h\|_W \qquad [1.31]$$

$$ii) \|p - p_h\|_Q \le \frac{M_a}{\beta^*} \|v - v_h\|_W + \left(1 + \frac{M_b}{\beta^*}\right) \inf_{q_h \in Q_h} \|p - q_h\|_Q \qquad [1.32]$$

$$iii) \|v - v_h\|_W \le \left(1 + \frac{M_a}{\alpha^*}\right) \inf_{\overline{w}_h \in V_h(S)} \|v - \overline{w}_h\|_W + \frac{M_b}{\alpha^*} \inf_{q_h \in Q_h} \|p - q_h\|_Q \qquad [1.33]$$

i) There exist the linear, continuous operators $B_h : W_h \to Q'_h$ and $B^*_h : Q_h \to W'_h$ such that

$$\forall w_h \in W_h, \forall q_h \in Q_h, \quad b(w_h, q_h) = \langle B_h w_h, q_h \rangle_{Q'_h, Q_h} = \langle w_h, B^*_h q_h \rangle_{W_h, W'_h}.$$

Given [1.29], by using lemma 1.1, we find that the operator $B_h : V_h^\perp \to Q'_h$ is bijective and $\left\|B_h v_h^\perp\right\|_{Q'_h} \ge \beta^* \left\|v_h^\perp\right\|_{W_h}$ for all $v_h^\perp \in V_h^\perp$.

Consider an arbitrary $w_h \in W_h$. The map $q_h \in Q_h \to b(v - w_h, q_h)$ is an element of Q'_h. Since $B_h : V_h^\perp \to Q'_h$ is bijective, there exists a unique $z_h \in V_h^\perp$ such that $\langle B_h z_h, q_h \rangle_{Q'_h, Q_h} = b(v - w_h, q_h)$ for all $q_h \in Q_h$. However,

$$\begin{aligned} \|B_h z_h\|_{Q'_h} &= \sup_{q_h \in Q_h, q_h \neq 0} \frac{\langle B_h z_h, q_h \rangle_{Q'_h, Q_h}}{\|q_h\|_{Q_h}} = \sup_{q_h \in Q_h, q_h \neq 0} \frac{b(v - w_h, q_h)}{\|q_h\|_{Q_h}} \\ &\le \sup_{q_h \in Q_h, q_h \neq 0} \frac{M_b \|v - w_h\|_W \|q_h\|_Q}{\|q_h\|_{Q_h}} = M_b \|v - w_h\|_W . \end{aligned}$$

We have used that $\|q_h\|_{Q_h} = \|q_h\|_Q$ if $q_h \in Q_h$. Since $\|B_h z_h\|_{Q'_h} \ge \beta^* \|z_h\|_{W_h}$, we obtain

$$\|z_h\|_{W_h} \le \frac{M_b}{\beta^*} \|v - w_h\|_W . \qquad [1.34]$$

To each $w_h \in W_h$, we relate $\overline{w}_h = z_h + w_h \in W_h$. For all $q_h \in Q_h$, we have

$$\begin{aligned} b(\overline{w}_h, q_h) &= b(z_h, q_h) + b(w_h, q_h) = \langle B_h z_h, q_h \rangle_{Q'_h, Q_h} + b(w_h, q_h) \\ &= b(v - w_h, q_h) + b(w_h, q_h) = b(v, q_h) = \langle S, q_h \rangle_{Q', Q}. \end{aligned}$$

For the last equation, we used [1.11]. We thus obtain that $\overline{w}_h \in V_h(S)$. We have

$$\|v - \overline{w}_h\|_W = \|v - w_h - z_h\|_W \leq \|v - w_h\|_W + \|z_h\|_W .$$

Taking into account [1.34], we have

$$\|v - \overline{w}_h\|_W \leq \left(1 + \frac{M_b}{\beta^*}\right) \|v - w_h\|_W .$$

We define the map $R : W_h \to W_h$ by $R(w_h) = \overline{w}_h$. We thus have

$$\forall w_h \in W_h, \quad \|v - R(w_h)\|_W \leq \left(1 + \frac{M_b}{\beta^*}\right) \|v - w_h\|_W$$

which implies

$$\inf_{w_h \in W_h} \|v - R(w_h)\|_W \leq \left(1 + \frac{M_b}{\beta^*}\right) \inf_{w_h \in W_h} \|v - w_h\|_W .$$

However, we have seen that $R(w_h) \in V_h(S)$, thus

$$\inf_{\overline{w}_h \in V_h(S)} \|v - \overline{w}_h\|_W \leq \inf_{w_h \in W_h} \|v - R(w_h)\|_W$$

which gives [1.31].

ii) We replace w with $w_h \in W_h$ in [1.10] and we subtract [1.24]; therefore,

$$\begin{aligned} & a(v - v_h, w_h) + b(w_h, p - p_h) = 0, \quad \forall w_h \in W_h \\ \Leftrightarrow\ & b(w_h, p_h) = a(v - v_h, w_h) + b(w_h, p), \quad \forall w_h \in W_h \\ \Leftrightarrow\ & b(w_h, p_h - q_h) = a(v - v_h, w_h) + b(w_h, p - q_h), \quad \forall w_h \in W_h, \forall q_h \in Q_h. \end{aligned}$$

Condition [1.29] is equivalent to

$$\forall q_h \in Q_h, \qquad \sup_{w_h \in W_h, w_h \neq 0} \frac{b(w_h, q_h)}{\|w_h\|_W} \geq \beta^* \|q_h\|_Q .$$

Since $p_h \in Q_h$,

$$\forall q_h \in Q_h, \qquad \sup_{w_h \in W_h, w_h \neq 0} \frac{b(w_h, p_h - q_h)}{\|w_h\|_W} \geq \beta^* \|p_h - q_h\|_Q .$$

Using the continuity of a and b, we obtain

$$\begin{aligned} |b(w_h, p_h - q_h)| &= |a(v - v_h, w_h) + b(w_h, p - q_h)| \\ &\leq M_a \|v - v_h\|_W \|w_h\|_W + M_b \|w_h\|_W \|p - q_h\|_Q \end{aligned}$$

from which

$$\begin{aligned} \beta^* \|p_h - q_h\|_Q &\leq \sup_{w_h \in W_h, w_h \neq 0} \frac{b(w_h, p_h - q_h)}{\|w_h\|_W} \\ &\leq \sup_{w_h \in W_h, w_h \neq 0} \frac{M_a \|v - v_h\|_W \|w_h\|_W + M_b \|w_h\|_W \|p - q_h\|_Q}{\|w_h\|_W} \\ &= M_a \|v - v_h\|_W + M_b \|p - q_h\|_Q . \end{aligned}$$

However, $\|p - p_h\|_Q \leq \|p - q_h\|_Q + \|q_h - p_h\|_Q$, thus

$$\begin{aligned} \|p - p_h\|_Q &\leq \|p - q_h\|_Q + \frac{M_a}{\beta^*} \|v - v_h\|_W + \frac{M_b}{\beta^*} \|p - q_h\|_Q \\ &= \frac{M_a}{\beta^*} \|v - v_h\|_W + \left(1 + \frac{M_b}{\beta^*}\right) \|p - q_h\|_Q . \end{aligned}$$

Since q_h is arbitrary, we can use $\inf_{q_h \in Q_h}$ and we obtain [1.32].

iii) Consider an arbitrary $\overline{w}_h \in V_h(S)$, thus $b(\overline{w}_h, q_h) = \langle S, q_h \rangle_{Q',Q}$ for all $q_h \in Q_h$. Thanks to [1.25], we have $v_h \in V_h(S)$, thus $b(v_h - \overline{w}_h, q_h) = 0$, for all $q_h \in Q_h$.

By replacing w with $v_h - \overline{w}_h$ in [1.10], we obtain

$$a(v, v_h - \overline{w}_h) + b(v_h - \overline{w}_h, p) = \langle f, v_h - \overline{w}_h \rangle_{W',W}$$

and by replacing w_h with $v_h - \overline{w}_h$ in [1.24], we obtain

$$a(v_h, v_h - \overline{w}_h) + b(v_h - \overline{w}_h, p_h) = \langle f, v_h - \overline{w}_h \rangle_{W',W}.$$

By subtraction, we have

$$a(v - v_h, v_h - \overline{w}_h) + b(v_h - \overline{w}_h, p - p_h) = 0$$

which implies

$$\begin{aligned} & a(v, v_h - \overline{w}_h) + b(v_h - \overline{w}_h, p - p_h) = a(v_h, v_h - \overline{w}_h) \\ \Leftrightarrow \; & a(v - \overline{w}_h, v_h - \overline{w}_h) + b(v_h - \overline{w}_h, p - p_h) = a(v_h - \overline{w}_h, v_h - \overline{w}_h). \end{aligned}$$

We have seen that $b(v_h - \overline{w}_h, q_h) = 0$, for all $q_h \in Q_h$ and since $p_h \in Q_h$, $b(v_h - \overline{w}_h, p - p_h) = b(v_h - \overline{w}_h, p) = b(v_h - \overline{w}_h, p - q_h)$. Finally, we obtain

$$a(v_h - \overline{w}_h, v_h - \overline{w}_h) = a(v - \overline{w}_h, v_h - \overline{w}_h) + b(v_h - \overline{w}_h, p - q_h).$$

We have $v_h - \overline{w}_h \in V_h$ and using the V_h-ellipticity of a and the continuity of a and b, we arrive at

$$\begin{aligned} & \alpha^* \|v_h - \overline{w}_h\|_W^2 \leq a(v_h - \overline{w}_h, v_h - \overline{w}_h) \\ & \leq M_a \|v - \overline{w}_h\|_W \|v_h - \overline{w}_h\|_W + M_b \|v_h - \overline{w}_h\|_W \|p - q_h\|_Q \\ \Leftrightarrow \; & \alpha^* \|v_h - \overline{w}_h\|_W \leq M_a \|v - \overline{w}_h\|_W + M_b \|p - q_h\|_Q . \end{aligned}$$

The triangle inequality gives $\|v - v_h\|_W \leq \|v - \overline{w}_h\|_W + \|v_h - \overline{w}_h\|_W$ and we obtain

$$\|v - v_h\|_W \leq \|v - \overline{w}_h\|_W + \frac{M_a}{\alpha^*} \|v - \overline{w}_h\|_W + \frac{M_b}{\alpha^*} \|p - q_h\|_Q .$$

This inequality is valid for all $\overline{w}_h \in V_h(S)$ and $q_h \in Q_h$, which implies [1.33].

We will now demonstrate [1.30]. Taking account of [1.33] and [1.31], we have

$$\|v - v_h\|_W \le \left(1 + \frac{M_a}{\alpha^*}\right)\left(1 + \frac{M_b}{\beta^*}\right) \inf_{w_h \in W_h} \|v - w_h\|_W + \frac{M_b}{\alpha^*} \inf_{q_h \in Q_h} \|p - q_h\|_Q .$$

Using [1.32] and the inequality above, we have

$$\begin{aligned}
\|p - p_h\|_Q \le & \frac{M_a}{\beta^*}\left(1 + \frac{M_a}{\alpha^*}\right)\left(1 + \frac{M_b}{\beta^*}\right) \inf_{w_h \in W_h} \|v - w_h\|_W \\
& + \frac{M_a}{\beta^*}\frac{M_b}{\alpha^*} \inf_{q_h \in Q_h} \|p - q_h\|_Q + \left(1 + \frac{M_b}{\beta^*}\right) \inf_{q_h \in Q_h} \|p - q_h\|_Q \\
= & \frac{M_a}{\beta^*}\left(1 + \frac{M_a}{\alpha^*}\right)\left(1 + \frac{M_b}{\beta^*}\right) \inf_{w_h \in W_h} \|v - w_h\|_W \\
& + \left(\frac{M_a}{\beta^*}\frac{M_b}{\alpha^*} + 1 + \frac{M_b}{\beta^*}\right) \inf_{q_h \in Q_h} \|p - q_h\|_Q .
\end{aligned}$$

Finally, we have

$$\begin{aligned}
& \|v - v_h\|_W + \|p - p_h\|_Q \\
& \le \left(1 + \frac{M_a}{\beta^*}\right)\left(1 + \frac{M_a}{\alpha^*}\right)\left(1 + \frac{M_b}{\beta^*}\right) \inf_{w_h \in W_h} \|v - w_h\|_W \\
& + \left(\frac{M_b}{\alpha^*} + \frac{M_a}{\beta^*}\frac{M_b}{\alpha^*} + 1 + \frac{M_b}{\beta^*}\right) \inf_{q_h \in Q_h} \|p - q_h\|_Q
\end{aligned}$$

which gives [1.30]. □

1.5. Mixed finite elements for the Stokes equations

We use the same notation as in section 1.3, in particular:

$$W = (H_0^1(\Omega))^2, \quad Q = L_0^2(\Omega), \quad V = \{\mathbf{w} \in (H_0^1(\Omega))^2;\ \nabla \cdot \mathbf{w} = 0\}.$$

Let $W_h \subset W$ and $Q_h \subset Q$ be two finite-dimension vector spaces and

$$V_h = \{w_h \in W_h;\ \forall q_h \in Q_h,\ b(w_h, q_h) = 0\}.$$

In general, V_h is not a subspace of V. We have seen that $a : W \times W \to \mathbb{R}$ is continuous; thus, $a : W_h \times W_h \to \mathbb{R}$ will be as well.

For the Stokes problem, we saw that a is W-elliptic, not just V-elliptic. Since $V_h \subset W$, a satisfies condition [1.28] with $\alpha^* = \alpha$ for any choice of $W_h \subset W$ and $Q_h \subset Q$! The map $b : W_h \times Q_h \to \mathbb{R}$ will be continuous, being the restriction of the continuous map $b : W \times Q \to \mathbb{R}$. In order to apply theorem 1.3, it remains for us to show that b satisfies the discrete inf-sup condition [1.29]. We will see that it is vital to choose the pair of finite-dimension vector spaces W_h, Q_h.

We will use the following result:

THEOREM 1.4.– *[FOR 77] We assume that $b : W \times Q \to \mathbb{R}$ satisfies the continuous inf-sup condition [1.9] and $W_h \subset W$, $Q_h \subset Q$ are two finite-dimension vector spaces. We assume that there exists a linear, continuous operator $\Pi_h : W \to W_h$, such that*

$$\forall w \in W, \forall q_h \in Q_h, \quad b(w - \Pi_h(w), q_h) = 0$$

and $\|\Pi_h\|_{\mathcal{L}(W,W_h)} \leq C$ where C is a constant independent of h. Hence, the discrete inf-sup condition [1.29] holds.

DEMONSTRATION 1.6.– The continuous inf-sup condition is equivalent to

$$\forall q \in Q, \quad \sup_{w \in W, w \neq 0} \frac{b(w,q)}{\|w\|_W} \geq \beta \|q\|_Q$$

and in particular for all $q_h \in Q_h$

$$\sup_{w \in W, w \neq 0} \frac{b(w,q_h)}{\|w\|_W} \geq \beta \|q_h\|_Q .$$

Using the hypotheses $b(w - \Pi_h(w), q_h) = 0$ and $\|\Pi_h(w)\|_W \leq C \|w\|_W$ for all $q_h \in Q_h$ and all $w \in W$, we have

$$\sup_{w \in W, w \neq 0} \frac{b(w,q_h)}{\|w\|_W} = \sup_{w \in W, w \neq 0} \frac{b(\Pi_h(w),q_h)}{\|w\|_W} \leq C \sup_{w \in W, w \neq 0} \frac{b(\Pi_h(w),q_h)}{\|\Pi_h(w)\|_W}.$$

Since $\Pi_h(W) \subset W_h$,

$$\sup_{w \in W, w \neq 0} \frac{b(\Pi_h(w),q_h)}{\|\Pi_h(w)\|_W} \leq \sup_{w_h \in W_h, w_h \neq 0} \frac{b(w_h,q_h)}{\|w_h\|_W}$$

and we obtain

$$\forall q_h \in Q_h, \quad \beta \|q_h\|_Q \leq C \sup_{w_h \in W_h, w_h \neq 0} \frac{b(w_h, q_h)}{\|w_h\|_W}$$

which is equivalent to the discrete inf-sup condition with $\beta^* = \beta/C$. □

We will present the finite element ($\mathbb{P}_1$ + *bubble*/$\mathbb{P}_1$) or MINI, introduced in [ARN 84].

DEFINITION 1.1.– *Let $\Omega \subset \mathbb{R}^2$ be an open, bounded, connected, polygonal set.* A triangulation (or mesh) *of Ω is a set $\mathcal{T}_h$ of triangles such that $\overline{\Omega} = \cup_{T \in \mathcal{T}_h} T$ with the following properties: if T_i, T_j are two distinct triangles, then $T_i \cap T_j$ is either the empty set, or a vertex, or a side. We consider a triangle as a closed set.*

For each $T \in \mathcal{T}_h$, we denote the maximum Euclidean distance between two points of T with h_T. We define $h = \max_{T \in \mathcal{T}_h} h_T$. Furthermore, for each $T \in \mathcal{T}_h$, we denote with ρ_T the maximum diameter of the circles contained in T.

DEFINITION 1.2.– *We say that $(\mathcal{T}_h)_{h>0}$ is a* regular *family of triangulations of $\overline{\Omega}$ if:*

$$h \to 0 \qquad [1.35]$$

$$\exists \sigma \geq 1, \ \forall h > 0, \ \forall T \in \mathcal{T}_h, \ \frac{h_T}{\rho_T} \leq \sigma. \qquad [1.36]$$

DEFINITION 1.3.– *We say that $(\mathcal{T}_h)_{h>0}$ is a* uniformly regular *family of triangulations of $\overline{\Omega}$ if it is regular and:*

$$\exists \sigma' \in]0, 1[, \ \forall h > 0, \ \forall T \in \mathcal{T}_h, \ \sigma' h \leq h_T. \qquad [1.37]$$

Let T be a triangle with vertices A, B, C at coordinates (x_1^A, x_2^A), (x_1^B, x_2^B), (x_1^C, x_2^C). We use G to denote the barycenter, *i.e.* the point with coordinates $x_1^G = \frac{x_1^A + x_1^B + x_1^C}{3}$ and $x_2^G = \frac{x_2^A + x_2^B + x_2^C}{3}$. For a point $\mathbf{x} = (x_1, x_2) \in \mathbb{R}^2$, we introduce the barycentric coordinates $\lambda_1(\mathbf{x})$, $\lambda_2(\mathbf{x})$, $\lambda_3(\mathbf{x})$, as the solution of the linear system

$$\begin{pmatrix} x_1^A & x_1^B & x_1^C \\ x_2^A & x_2^B & x_2^C \\ 1 & 1 & 1 \end{pmatrix} \begin{pmatrix} \lambda_1(\mathbf{x}) \\ \lambda_2(\mathbf{x}) \\ \lambda_3(\mathbf{x}) \end{pmatrix} = \begin{pmatrix} x_1 \\ x_2 \\ 1 \end{pmatrix}.$$

We have $\lambda_i(\mathbf{x}) > 0$ if $\mathbf{x} \in int(T)$ and $\lambda_i(\mathbf{x}) = 0$ if $\mathbf{x} \in \partial T$. We will define the bubble function $b_T : \overline{\Omega} \to \mathbb{R}$ as $b_T(\mathbf{x}) = \lambda_1(\mathbf{x})\lambda_2(\mathbf{x})\lambda_3(\mathbf{x})$ if $\mathbf{x} \in T$ and $b_T(\mathbf{x}) = 0$ if not. Let the sets of functions be

$$\mathbb{P}_1(T) = \{r : T \to \mathbb{R};\ r(x_1, x_2) = a\,x_1 + b\,x_2 + c;\ a, b, c \in \mathbb{R}\}$$

$$(\mathbb{P}_1 + b)(T) = \{r : T \to \mathbb{R};\ r(x_1, x_2) = a\,x_1 + b\,x_2 + c + d\,b_T(x_1, x_2);$$
$$a, b, c, d \in \mathbb{R}\}.$$

We define the finite element $\mathbb{P}_1$ + *bubble* or $\mathbb{P}_1 + b$, the triplet

$$(T, \{A, B, C, G\}, (\mathbb{P}_1 + b)(T))\,.$$

The finite element $\mathbb{P}_1$ is the triplet $(T, \{A, B, C\}, \mathbb{P}_1(T))$.

We will introduce the finite-dimension spaces

$$Q_h = \{q_h \in C^0(\overline{\Omega});\ \forall T \in \mathcal{T}_h,\ q_{h|T} \in \mathbb{P}_1(T),\ \int_\Omega q_h d\mathbf{x} = 0\}$$

$$W_h = \{w_h = (w_h^1, w_h^2) \in (C^0(\overline{\Omega}))^2;\ w_h = 0 \text{ on } \partial\Omega,$$
$$\forall T \in \mathcal{T}_h,\ w_{h|T}^i \in (\mathbb{P}_1 + b)(T),\ i = 1, 2\}.$$

Since $C^0(\overline{\Omega}) \subset L^2(\Omega)$, $Q_h \subset Q$. For $\Omega \subset \mathbb{R}^2$, $C^0(\overline{\Omega})$ is not a subspace of $H^1(\Omega)$. We will apply Theorem 1.4-3, p. 27, [RAV 98]: if $w_{h|T}^i \in (\mathbb{P}_1 + b)(T) \subset H^1(int(T))$ for every $T \in \mathcal{T}_h$ and $w_h^i \in C^0(\overline{\Omega})$, then $w_h^i \in H^1(\Omega)$. Taking account of the boundary conditions, we obtain $W_h \subset W$.

It remains to construct an operator that satisifies theorem 1.4.

For a function $w \in C^0(\overline{\Omega})$, the value of w at a point is well-defined. We can construct $w_h \in W_h$ by interpolation, $w_h = P$ over each $T \in \mathcal{T}_h$ such that $P \in (\mathbb{P}_1 + b)(T)$, $P(A) = w(A)$, $P(B) = w(B)$, $P(C) = w(C)$, $P(G) = w(G)$, where A, B, C and G are the vertices and the barycenter of T, respectively.

For $\Omega \subset \mathbb{R}^2$, $H^1(\Omega)$ is not a subspace of $C^0(\overline{\Omega})$; thus, for $w \in H^1(\Omega)$, we cannot construct the operator from interpolation as above. We will use the following result:

THEOREM 1.5.– *[CLÉ 75] Let* $\Omega \subset \mathbb{R}^2$ *be an open, bounded, connected, polygonal set and* $(\mathcal{T}_h)_{h>0}$ *be a regular family of triangulations of* $\overline{\Omega}$. *There exists a linear, continuous operator* $\pi_h^0 : H_0^1(\Omega) \rightarrow X_{h0}$ *such that*

$$\exists C_1 > 0, \forall w \in H_0^1(\Omega),\ \left\|\pi_h^0(w)\right\|_{1,\Omega} \leq C_1 \|w\|_{1,\Omega} \qquad [1.38]$$

$$\exists C_2 > 0, \forall w \in H_0^1(\Omega),\ \left\|w - \pi_h^0(w)\right\|_{0,\Omega} \leq C_2\, h\, \|w\|_{1,\Omega} \qquad [1.39]$$

where

$$X_{h0} = \{w_h \in C^0(\overline{\Omega});\ \forall T \in \mathcal{T}_h,\ w_{h|T} \in \mathbb{P}_1(T),\ w_h = 0 \text{ on } \partial\Omega\}.$$

THEOREM 1.6.– *Let* $\Omega \subset \mathbb{R}^2$ *be an open, bounded, connected, polygonal set and* $(\mathcal{T}_h)_{h>0}$ *be a uniformly regular family of triangulations of* $\overline{\Omega}$. *Hence, for* W_h *and* Q_h *defined previously, using the finite elements* $\mathbb{P}_1 + b/\mathbb{P}_1$, *the discrete inf-sup condition [1.29] holds with* β^* *independent of h.*

DEMONSTRATION 1.7.– Let

$$\mathbb{B}(\overline{\Omega}) = \{\sum_{T\in\mathcal{T}_h} \gamma_T\, b_T(\mathbf{x});\ \gamma_T \in \mathbb{R}\}.$$

We will define $\pi_h^1 : H_0^1(\Omega) \rightarrow \mathbb{B}(\overline{\Omega})$ by

$$\pi_h^1(w) = \sum_{T\in\mathcal{T}_h} \alpha_T(w) b_T, \quad \alpha_T(w) = \frac{\int_T w\, d\mathbf{x}}{\int_T b_T(\mathbf{x})\, d\mathbf{x}}.$$

Since $b_T \in H_0^1(\Omega)$, $\pi_h^1(w) \in H_0^1(\Omega)$. First, we will show that there exists C_3 independent of h and

$$\forall w \in H_0^1(\Omega),\ \left\|\pi_h^1(w)\right\|_{0,\Omega} \leq C_3 \|w\|_{0,\Omega}\,. \qquad [1.40]$$

We use $\hat{T}$ to denote the reference triangle with the vertices $(0,1)$, $(1,0)$ and $(0,0)$. We use $\hat{\lambda}_i(\hat{\mathbf{x}})$ to denote the barycentric coordinates of $\hat{\mathbf{x}}$ with respect to the vertices of $\hat{T}$. There exists an affine map

$$F : \mathbb{R}^2 \rightarrow \mathbb{R}^2, \quad F(\hat{\mathbf{x}}) = B\hat{\mathbf{x}} + b$$

where B is a real, invertible 2×2 matrix and $b \in \mathbb{R}^2$ such that $F(\hat{T}) = T$. Since $\hat{\lambda}_i(\hat{\mathbf{x}}) = \lambda_i(\mathbf{x})$, where $\mathbf{x} = F(\hat{\mathbf{x}})$ (see, for example [DAU 88], vol. 9, chap. XII,

p. 750), $\hat{b}_{\hat{T}}(\hat{\mathbf{x}}) = b_T(\mathbf{x})$. We have used $\hat{b}_{\hat{T}}$ to denote the bubble function associated with the triangle $\hat{T}$. We have

$$\left\|\pi_h^1(w)\right\|_{0,\Omega}^2 = \sum_{T\in\mathcal{T}_h} \|\alpha_T(w)b_T\|_{0,T}^2 = \sum_{T\in\mathcal{T}_h} |\alpha_T(w)|^2\,\|b_T\|_{0,T}^2$$

$$= \sum_{T\in\mathcal{T}_h} \left|\frac{\int_T w\,d\mathbf{x}}{\int_T b_T(\mathbf{x})\,d\mathbf{x}}\right|^2 \|b_T\|_{0,T}^2 .$$

We will carry out a change in coordinates for the integrals concerning b_T.

$$\|b_T\|_{0,T}^2 = \int_T b_T^2(\mathbf{x})\,d\mathbf{x} = \int_{\hat{T}} \hat{b}_{\hat{T}}^2(\hat{\mathbf{x}})\det(B)\,d\hat{\mathbf{x}} = \det(B)\int_{\hat{T}} \hat{b}_{\hat{T}}^2(\hat{\mathbf{x}})\,d\hat{\mathbf{x}}$$

because $\det(B) > 0$ is a constant. Similarly, we have

$$\int_T b_T(\mathbf{x})\,d\mathbf{x} = \det(B)\int_{\hat{T}} \hat{b}_{\hat{T}}(\hat{\mathbf{x}})\,d\hat{\mathbf{x}}.$$

We use $|T|$ to denote the area of triangle T. The Cauchy-Schwarz inequality gives

$$\left|\int_T w\,d\mathbf{x}\right|^2 \le |T|\,\|w\|_{0,T}^2 .$$

We also have $|T| = \det(B)|\hat{T}|$. We obtain

$$\left|\frac{\int_T w\,d\mathbf{x}}{\int_T b_T(\mathbf{x})\,d\mathbf{x}}\right|^2 \|b_T\|_{0,T}^2 \le \frac{|T|\,\|w\|_{0,T}^2}{\left(\det(B)\int_{\hat{T}} \hat{b}_{\hat{T}}(\hat{\mathbf{x}})\,d\hat{\mathbf{x}}\right)^2}\det(B)\int_{\hat{T}} \hat{b}_{\hat{T}}^2(\hat{\mathbf{x}})\,d\hat{\mathbf{x}}$$

$$= \|w\|_{0,T}^2\,\frac{|\hat{T}|\int_{\hat{T}} \hat{b}_{\hat{T}}^2(\hat{\mathbf{x}})\,d\hat{\mathbf{x}}}{\left(\int_{\hat{T}} \hat{b}_{\hat{T}}(\hat{\mathbf{x}})\,d\hat{\mathbf{x}}\right)^2}.$$

The last fraction is independent of w and h and we denote it with C_3. Finally, we have

$$\left\|\pi_h^1(w)\right\|_{0,\Omega}^2 \le C_3 \sum_{T\in\mathcal{T}_h} \|w\|_{0,T}^2 = C_3\,\|w\|_{0,\Omega}^2$$

from which [1.40].

Now we will show that there exists C_4 independent of h and

$$\forall w \in H_0^1(\Omega),\ \left\|\pi_h^1(w)\right\|_{1,\Omega} \leq C_4 \frac{1}{h} \left\|\pi_h^1(w)\right\|_{0,\Omega}. \qquad [1.41]$$

We have

$$\left\|\pi_h^1(w)\right\|_{1,\Omega}^2 = \sum_{T \in \mathcal{T}_h} ||\alpha_T(w) b_T||_{1,T}^2 = \sum_{T \in \mathcal{T}_h} |\alpha_T(w)|^2 \, ||b_T||_{1,T}^2$$
$$= \sum_{T \in \mathcal{T}_h} |\alpha_T(w)|^2 \, ||b_T||_{0,T}^2 \frac{||b_T||_{1,T}^2}{||b_T||_{0,T}^2}.$$

We will demonstrate that there exists C_4 independent of h such that $\frac{||b_T||_{1,T}^2}{||b_T||_{0,T}^2} \leq \frac{C_4^2}{h^2}$. We have $||b_T||_{1,T}^2 = ||b_T||_{0,T}^2 + ||\nabla b_T||_{0,T}^2$. Using the change in coordinates, we obtain

$$||b_T||_{0,T}^2 = \det(B) \int_{\hat{T}} \hat{b}_{\hat{T}}^2(\hat{\mathbf{x}})\, d\hat{\mathbf{x}}.$$

Since $\det(B) = \frac{|T|}{|\hat{T}|} = 2|T|$ and $|T| \leq |\Omega|$, then $||b_T||_{0,T}$ is bounded by a constant independent of h. Thanks to Theorem 8, [DAU 88], vol. 6, chap. XII, p. 755, we have

$$||\nabla b_T||_{0,T} \leq C\, ||B|| \,\sqrt{\det(B)}\, \left\|\nabla \hat{b}_{\hat{T}}\right\|_{0,\hat{T}}$$

and using Theorem 9, [DAU 88], vol. 6, chap. XII, p. 757, we have $||B|| \leq \frac{\hat{h}}{\rho_T}$. Since $\det(B) \leq h_T^2$, we obtain

$$||\nabla b_T||_{0,T} \leq \frac{h_T}{\rho_T} C\hat{h} \left\|\nabla \hat{b}_{\hat{T}}\right\|_{0,\hat{T}}$$

and taking into account that the triangulation is regular, therefore $\frac{h_T}{\rho_T} \leq \sigma$, we find that $||\nabla b_T||_{0,T}$ is bounded by a constant independent of h. We deduce that $||b_T||_{1,T}$ is bounded by a constant independent of h. We have $\det(B) = 2|T|$

and the area of the triangle is larger than the area of the contained circle; thus, $|T| \geq \pi\rho_T^2$. We have

$$\|b_T\|_{0,T}^2 \geq 2\pi\rho_T^2 \int_{\hat{T}} \hat{b}_{\hat{T}}^2(\hat{\mathbf{x}})\, d\hat{\mathbf{x}}$$

but the triangulation is uniformly regular; thus, $\rho_T \geq \frac{\sigma'}{\sigma}h$. The result is that $\|b_T\|_{0,T}^2 \geq h^2 \times const$, which tells us that there exists a constant, C_4, independent of h such that $\frac{\|b_T\|_{1,T}^2}{\|b_T\|_{0,T}^2} \leq \frac{C_4^2}{h^2}$. We deduce that

$$\begin{aligned} \left\|\pi_h^1(w)\right\|_{1,\Omega}^2 &= \sum_{T\in\mathcal{T}_h} |\alpha_T(w)|^2 \, \|b_T\|_{0,T}^2 \, \frac{\|b_T\|_{1,T}^2}{\|b_T\|_{0,T}^2} \\ &\leq \frac{C_4^2}{h^2} \sum_{T\in\mathcal{T}_h} |\alpha_T(w)|^2 \, \|b_T\|_{0,T}^2 = \frac{C_4^2}{h^2} \left\|\pi_h^1(w)\right\|_{0,\Omega}^2 \end{aligned}$$

which gives us [1.41].

We will now introduce Π_h, an operator that satisfies theorem 1.4 for the Stokes equations. For $\mathbf{w} = (w_1, w_2) \in W$, we will denote $\Pi_h^0(\mathbf{w}) = \left(\pi_h^0(w_1), \pi_h^0(w_2)\right)$ and $\Pi_h^1(\mathbf{w}) = \left(\pi_h^1(w_1), \pi_h^1(w_2)\right)$.

Let $\Pi_h : W \to W_h$, defined by

$$\Pi_h(\mathbf{w}) = \Pi_h^0(\mathbf{w}) + \Pi_h^1\left(\mathbf{w} - \Pi_h^0(\mathbf{w})\right).$$

We will start with the continuity of Π_h. Let $\mathbf{w} \in W$. The triangle inequality gives us

$$\|\Pi_h(\mathbf{w})\|_{1,\Omega} \leq \left\|\Pi_h^0(\mathbf{w})\right\|_{1,\Omega} + \left\|\Pi_h^1\left(\mathbf{w} - \Pi_h^0(\mathbf{w})\right)\right\|_{1,\Omega}$$

and first using [1.38], which represents the continuity of Π_h^0, followed by [1.41], we deduce

$$\begin{aligned} &\left\|\Pi_h^0(\mathbf{w})\right\|_{1,\Omega} + \left\|\Pi_h^1\left(\mathbf{w} - \Pi_h^0(\mathbf{w})\right)\right\|_{1,\Omega} \\ &\leq C_1 \|\mathbf{w}\|_{1,\Omega} + \frac{C_4}{h} \left\|\Pi_h^1\left(\mathbf{w} - \Pi_h^0(\mathbf{w})\right)\right\|_{0,\Omega}. \end{aligned}$$

In what follows, we use [1.40] and [1.39]

$$\begin{aligned}
&C_1 \|\mathbf{w}\|_{1,\Omega} + \frac{C_4}{h} \left\|\Pi_h^1\left(\mathbf{w} - \Pi_h^0(\mathbf{w})\right)\right\|_{0,\Omega} \\
&\leq C_1 \|\mathbf{w}\|_{1,\Omega} + \frac{C_4}{h} C_3 \left\|\mathbf{w} - \Pi_h^0(\mathbf{w})\right\|_{0,\Omega} \\
&\leq C_1 \|\mathbf{w}\|_{1,\Omega} + \frac{C_4}{h} C_3 C_2 h \|\mathbf{w}\|_{1,\Omega} .
\end{aligned}$$

Finally, we have the continuity of Π_h

$$\forall \mathbf{w} \in W, \quad \|\Pi_h(\mathbf{w})\|_{1,\Omega} \leq (C_1 + C_2 C_3 C_4) \|\mathbf{w}\|_{1,\Omega} .$$

The last step is to show that

$$\forall \mathbf{w} \in W, \forall q_h \in Q_h, \quad b(\mathbf{w} - \Pi_h(\mathbf{w}), q_h) = 0.$$

For the Stokes equations, we have

$$\begin{aligned}
b(\mathbf{w} - \Pi_h(\mathbf{w}), q_h) &= -\int_\Omega \nabla \cdot (\mathbf{w} - \Pi_h(\mathbf{w})) q_h \, d\mathbf{x} \\
&= \int_\Omega (\mathbf{w} - \Pi_h(\mathbf{w})) \cdot \nabla q_h \, d\mathbf{x} - \int_{\partial\Omega} (\mathbf{w} - \Pi_h(\mathbf{w})) q_h \cdot \mathbf{n} \, ds.
\end{aligned}$$

However, $\pi_h^0(w_i) \in X_{h0} \subset H_0^1(\Omega)$ and $\pi_h^1(w_i) \in \mathbb{B}(\overline{\Omega}) \subset H_0^1(\Omega)$; therefore, the integral over $\partial\Omega$ is zero. We have

$$\int_\Omega (\mathbf{w} - \Pi_h(\mathbf{w})) \cdot \nabla q_h \, d\mathbf{x} = \sum_{T \in \mathcal{T}_h} \int_T (\mathbf{w} - \Pi_h(\mathbf{w})) \cdot \nabla q_h \, d\mathbf{x}$$

and over each triangle $T \in \mathcal{T}_h$, $\nabla q_h = (q_1, q_2)^T$ and $q_1, q_2 \in \mathbb{R}$ are two constants, because $q_{h|T} \in \mathbb{P}_1(T)$. We have

$$\int_T (\mathbf{w} - \Pi_h(\mathbf{w})) \cdot \nabla q_h \, d\mathbf{x} = \sum_{i=1}^{2} q_i \int_T (w_i - \pi_h(w_i)) \, d\mathbf{x}$$

with $\pi_h(w_i) = \pi_h^0(w_i) + \pi_h^1\left(w_i - \pi_h^0(w_i)\right)$. Thus, we deduce

$$\int_T (w_i - \pi_h(w_i))\, d\mathbf{x} = \int_T w_i - \pi_h^0(w_i) - \pi_h^1\left(w_i - \pi_h^0(w_i)\right)\, d\mathbf{x}$$
$$= \int_T \overline{w}_i - \pi_h^1(\overline{w}_i)\, d\mathbf{x}$$

where $\overline{w}_i = w_i - \pi_h^0(w_i)$. Using the definition of π_h^1, we have

$$\int_T \overline{w}_i - \pi_h^1(\overline{w}_i)\, d\mathbf{x} = \int_T \overline{w}_i\, d\mathbf{x} - \int_T \pi_h^1(\overline{w}_i)\, d\mathbf{x}$$
$$= \int_T \overline{w}_i\, d\mathbf{x} - \int_T \alpha_T(\overline{w}_i) b_T(\mathbf{x})\, d\mathbf{x} = \int_T \overline{w}_i\, d\mathbf{x} - \int_T \frac{\int_T \overline{w}_i\, d\mathbf{x}}{\int_T b_T(\mathbf{x})\, d\mathbf{x}} b_T(\mathbf{x})\, d\mathbf{x}$$
$$= \int_T \overline{w}_i\, d\mathbf{x} - \frac{\int_T \overline{w}_i\, d\mathbf{x}}{\int_T b_T(\mathbf{x})\, d\mathbf{x}} \int_T b_T(\mathbf{x})\, d\mathbf{x} = \int_T \overline{w}_i\, d\mathbf{x} - \int_T \overline{w}_i\, d\mathbf{x} = 0.$$

We can now apply theorem 1.4 for the Stokes equations and the demonstration is complete. □

REMARK 1.5.– *To demonstrate [1.40] and [1.41], we have adapted the results of section 4.3 [SCH 13]. For the continuity of Π_h, we followed [BRA 01], Theorem 7.2, p. 165.*

REMARK 1.6.– *If, in addition, Ω is convex, the solution of the Stokes equation is more regular $\mathbf{v} \in (H^2(\Omega))^2$ and $p \in H^1(\Omega)$ (see [KEL 76]). We have $H^2(\Omega) \subset C^0(\overline{\Omega})$ in dimension two and there exists the interpolation operator $R_h : W \to X_{h0} \times X_{h0} \subset W_h$ such that*

$$\|\mathbf{v} - R_h(\mathbf{v})\|_{1,\Omega} \leq C_1\, h\, \|\mathbf{v}\|_{2,\Omega}$$

where h is the size of the triangulation. The Clément interpolation operator of theorem 1.5 satisfies

$$\left\|p - \pi_h^0(p)\right\|_{0,\Omega} \leq C_2\, h\, \|p\|_{1,\Omega}\,.$$

Hence,

$$\inf_{w_h \in W_h} \|\mathbf{v} - w_h\|_W \leq \|\mathbf{v} - R_h(\mathbf{v})\|_{1,\Omega} \text{ and } \inf_{q_h \in Q_h} \|p - q_h\|_Q \leq \left\|p - \pi_h^0(p)\right\|_{0,\Omega}$$

and using theorem 1.3, we deduce that

$$\|\mathbf{v} - \mathbf{v}_h\|_{1,\Omega} + \|p - p_h\|_{0,\Omega} \leq h\,C\left(\|\mathbf{v}\|_{2,\Omega} + \|p\|_{1,\Omega}\right). \qquad [1.42]$$

We will finish this section with the presentation of the finite elements $\mathbb{P}_2/\mathbb{P}_1$. Let T be a triangle with vertices A, B, C. We denote the middle of each side with D, E, F respectively and

$$(\mathbb{P}_2)(T) = \{r : T \to \mathbb{R};\ r(x_1, x_2) = a\,x_1^2 + b\,x_2^2 + c\,x_1x_2 + d\,x_1 + e\,x_2 + f;$$
$$a, b, c, d, e, f \in \mathbb{R}\}.$$

The finite element $\mathbb{P}_2$ is the triplet $(T, \{A, B, C, D, E, F\}, \mathbb{P}_2(T))$. The pair $\mathbb{P}_2/\mathbb{P}_1$ satisfies the discrete inf-sup condition for the Stokes equations. Alternative options for approximating the Stokes equations in 2D and 3D can be found in [GIR 86] and [PIR 89].

1.6. Boundary conditions for the Stokes equations

1.6.1. *Non-homogeneous Dirichlet condition*

We assume that $\Omega \subset \mathbb{R}^2$ is an open, non-empty, bounded, connected set and its boundary $\partial\Omega$ is Lipschitz. We are looking for the velocity $\mathbf{v} = (v_1, v_2) : \overline{\Omega} \to \mathbb{R}^2$ and the pressure $p : \overline{\Omega} \to \mathbb{R}$, such that

$$-2\mu\nabla \cdot \boldsymbol{\epsilon}(\mathbf{v}) + \nabla p = \mathbf{f}, \quad \text{in } \Omega, \qquad [1.43]$$

$$\nabla \cdot \mathbf{v} = 0, \quad \text{in } \Omega, \qquad [1.44]$$

$$\mathbf{v} = \mathbf{g}, \quad \text{on } \partial\Omega, \qquad [1.45]$$

where $\mu > 0$, $\mathbf{f} = (f_1, f_2) : \Omega \to \mathbb{R}^2$ and $\mathbf{g} : \partial\Omega \to \mathbb{R}^2$. We will specify the regularity later.

If $(\mathbf{v}, p)$ is a solution of [1.43]–[1.45], then $(\mathbf{v}, p + const)$ is also a solution. We say that the pressure is determined up to an additive constant. We can choose a representative pressure by setting $\int_\Omega p\,d\mathbf{x} = 0$. Making use of Green's formula for $\mathbf{v}$ and the constant function 1, using [1.45], we have

$$\int_\Omega (\nabla \cdot \mathbf{v})1\,d\mathbf{x} = \int_{\partial\Omega} (\mathbf{v} \cdot \mathbf{n})1\,ds - \int_\Omega \mathbf{v} \cdot \nabla 1\,d\mathbf{x} = \int_{\partial\Omega} (\mathbf{v} \cdot \mathbf{n})1\,ds = \int_{\partial\Omega} \mathbf{g} \cdot \mathbf{n}\,ds.$$

Taking account of [1.44], $\mathbf{g}$ must satisfy the condition

$$\int_{\partial\Omega} \mathbf{g} \cdot \mathbf{n}\, ds = 0 \qquad [1.46]$$

where $\mathbf{n}$ is the unit vector normal to the boundary, pointing outwards.

Let $\mathbf{w} : \overline{\Omega} \to \mathbb{R}^2$ such that $\mathbf{w} = 0$ on $\partial\Omega$. As for the homogeneous case, by multiplying [1.43] by $\mathbf{w}$, and integrating over Ω, applying Green's formula and using the symmetry of ϵ, we have

$$2\mu \int_{\Omega} \epsilon(\mathbf{v}) : \epsilon(\mathbf{w})\, d\mathbf{x} - \int_{\Omega} (\nabla \cdot \mathbf{w})\, p\, d\mathbf{x} = \int_{\Omega} \mathbf{f} \cdot \mathbf{w}\, d\mathbf{x}.$$

By multiplying [1.44] by $q : \overline{\Omega} \to \mathbb{R}$ satisfying $\int_{\Omega} q\, d\mathbf{x} = 0$ and integrating over Ω, we obtain

$$\int_{\Omega} (\nabla \cdot \mathbf{v})\, q\, d\mathbf{x} = 0.$$

The fact that $\int_{\Omega} q\, d\mathbf{x} = 0$ is not essential for the above equation. We use the notation

$$W = (H_0^1(\Omega))^2, \quad Q = L_0^2(\Omega) = \{q \in L^2(\Omega); \int_{\Omega} q\, d\mathbf{x} = 0\}.$$

We assume that $\mathbf{f} \in (L^2(\Omega))^2$ and $\mathbf{g} \in (H^{1/2}(\partial\Omega))^2$ such that [1.46] holds. For the definition of $H^{1/2}(\partial\Omega)$, see Appendix A.1. Find $\mathbf{v} \in (H^1(\Omega))^2$, $\mathbf{v} = \mathbf{g}$ on $\partial\Omega$ and $p \in Q$ such that

$$2\mu \int_{\Omega} \epsilon(\mathbf{v}) : \epsilon(\mathbf{w})\, d\mathbf{x} - \int_{\Omega} (\nabla \cdot \mathbf{w})\, p\, d\mathbf{x} = \int_{\Omega} \mathbf{f} \cdot \mathbf{w}\, d\mathbf{x}, \forall \mathbf{w} \in W \qquad [1.47]$$

$$-\int_{\Omega} (\nabla \cdot \mathbf{v})\, q\, d\mathbf{x} = 0, \ \forall q \in Q. \qquad [1.48]$$

We observe that $\mathbf{v} \notin W$; thus, we cannot directly apply the results obtained for the homogeneous Dirichlet boundary conditions. With the usual notation

$$a(\mathbf{v}, \mathbf{w}) = \int_{\Omega} 2\mu\, \epsilon(\mathbf{v}) : \epsilon(\mathbf{w})\, d\mathbf{x}, \quad b(\mathbf{v}, q) = -\int_{\Omega} (\nabla \cdot \mathbf{v})\, q\, d\mathbf{x}$$

we can re-write [1.47]–[1.48]

$$a(\mathbf{v}, \mathbf{w}) + b(\mathbf{w}, p) = \int_\Omega \mathbf{f} \cdot \mathbf{w} \, d\mathbf{x}, \quad \forall \mathbf{w} \in W$$

$$b(\mathbf{v}, q) = 0, \quad \forall q \in Q.$$

There exists $\mathbf{v}^g \in (H^1(\Omega))^2$ such that $\mathbf{v}^g = \mathbf{g}$ on $\partial\Omega$. The homogeneous Dirichlet problem: find $\mathbf{v}^0 \in W$, $p \in Q$ such that

$$a(\mathbf{v}^0, \mathbf{w}) + b(\mathbf{w}, p) = \int_\Omega \mathbf{f} \cdot \mathbf{w} \, d\mathbf{x} - a(\mathbf{v}^g, \mathbf{w}), \quad \forall \mathbf{w} \in W$$

$$b(\mathbf{v}^0, q) = -b(\mathbf{v}^g, q), \quad \forall q \in Q$$

has a unique solution. Hence, $\mathbf{v} = \mathbf{v}^0 + \mathbf{v}^g$ and p is the solution of the problem with non-homogeneous Dirichlet boundary conditions. In the previous sections, to obtain the discrete version, we replaced W by W_h and Q by Q_h in the mixed continuous problem, which gives: find $\mathbf{v}_h^0 \in W_h$ and $p_h \in Q_h$ such that

$$a(\mathbf{v}_h^0, \mathbf{w}_h) + b(\mathbf{w}_h, p_h) = \int_\Omega \mathbf{f} \cdot \mathbf{w}_h \, d\mathbf{x} - a(\mathbf{v}^g, \mathbf{w}_h), \quad \forall \mathbf{w}_h \in W_h$$

$$b(\mathbf{v}_h^0, q_h) = -b(\mathbf{v}^g, q_h), \quad \forall q_h \in Q_h.$$

This discrete problem has a unique solution, but $\mathbf{v}_h^0 + \mathbf{v}^g$ is not a finite element-type function. We must replace $\mathbf{v}^g$ with $\mathbf{v}_h^g$ that satisfies $\mathbf{v}_h^g = \mathbf{g}_h$ on the boundary where $\mathbf{g}_h$ is an approximation of $\mathbf{g}$. For more on this subject, see [BOF 13], starting with Remark 8.2.2, p. 464.

1.6.2. *Non-homogeneous Neumann condition*

We assume that $\Omega \subset \mathbb{R}^2$ is an open, non-empty, bounded, connected set and its boundary is Lipschitz $\partial\Omega = \overline{\Gamma}_D \cup \overline{\Gamma}_N$, $\Gamma_D \cap \Gamma_N = \emptyset$ and $leng(\Gamma_D) > 0$, $leng(\Gamma_N) > 0$. We have used $leng()$ to denote the curve length. Let $\mathbf{h} : \Gamma_N \to \mathbb{R}^2$ and denote the stress tensor with $\boldsymbol{\sigma} = -p\mathbf{I} + 2\mu\epsilon(\mathbf{v})$, where $\mathbf{I}$ is the unit matrix. We are looking for the velocity $\mathbf{v} = (v_1, v_2) : \overline{\Omega} \to \mathbb{R}^2$ and the pressure

$p : \overline{\Omega} \to \mathbb{R}$ satisfying [1.43], [1.44] and the boundary conditions

$$\mathbf{v} = 0, \quad \text{on } \Gamma_D, \qquad [1.49]$$

$$\sigma \mathbf{n} = \mathbf{h}, \quad \text{on } \Gamma_N. \qquad [1.50]$$

Equation [1.50] is also called the traction condition, because $\mathbf{h}$ has the dimensions of force per unit surface area in 3D. We remark that if $(\mathbf{v}, p)$ is a solution, then $(\mathbf{v}, p + const)$ no longer satisfies [1.50].

Let $\mathbf{w} : \overline{\Omega} \to \mathbb{R}^2$ such that $\mathbf{w} = 0$ on Γ_D. As for the homogeneous case, by multiplying [1.43] by $\mathbf{w}$, integrating over Ω, applying Green's formula and using the symmetry of $\boldsymbol{\epsilon}$, we have

$$2\mu \int_\Omega \boldsymbol{\epsilon}(\mathbf{v}) : \boldsymbol{\epsilon}(\mathbf{w})\, d\mathbf{x} - \int_\Omega (\nabla \cdot \mathbf{w})\, p\, d\mathbf{x} = \int_\Omega \mathbf{f} \cdot \mathbf{w}\, d\mathbf{x} + \int_{\Gamma_N} \sigma \mathbf{n} \cdot \mathbf{w}\, ds.$$

We can introduce the mixed formulation. This time we use the spaces

$$W = \{\mathbf{w} \in (H^1(\Omega))^2;\ \mathbf{w} = 0 \text{ on } \Gamma_D\}, \quad Q = L^2(\Omega).$$

We assume that $\mathbf{f} \in (L^2(\Omega))^2$ and $\mathbf{h} \in (L^2(\Gamma_N))^2$. Find $\mathbf{v} \in W$ and $p \in Q$ such that

$$2\mu \int_\Omega \boldsymbol{\epsilon}(\mathbf{v}) : \boldsymbol{\epsilon}(\mathbf{w})\, d\mathbf{x} - \int_\Omega (\nabla \cdot \mathbf{w})\, p\, d\mathbf{x}$$

$$= \int_\Omega \mathbf{f} \cdot \mathbf{w}\, d\mathbf{x} + \int_{\Gamma_N} \mathbf{h} \cdot \mathbf{w}\, ds, \quad \forall \mathbf{w} \in W \qquad [1.51]$$

$$-\int_\Omega (\nabla \cdot \mathbf{v})\, q\, d\mathbf{x} = 0, \quad \forall q \in Q. \qquad [1.52]$$

PROPOSITION 1.2.– *Problem [1.51]–[1.52] has a unique solution* $\mathbf{v} \in W$ *and* $p \in Q$.

With the usual notation, the W-ellipticity of a results from the Korn inequality in the case where we impose the homogeneous Dirichlet conditions only over a part of the boundary (see [CIA 04], Theorem 6.3-4, p. 292). We have seen that to show the continuous inf-sup condition, it is sufficient to prove that the divergence operator div : $W \to Q$, div $(\mathbf{w}) = \nabla \cdot \mathbf{w}$ is surjective. In the previous sections, we used the fact that div : $(H_0^1(\Omega))^2 \to L_0^2(\Omega)$ is surjective. We can also define div over $(H^1(\Omega))^2$ and div $\left((H^1(\Omega))^2\right) \subset L^2(\Omega)$. Since $(H_0^1(\Omega))^2 \subset W \subset (H^1(\Omega))^2$,

$$L_0^2(\Omega) \subset \text{div}\,(W) \subset L^2(\Omega)$$

and div (W) is a vector subspace. We can show that $L_0^2(\Omega) \neq \text{div}\,(W)$ and $L_0^2(\Omega)$ is a subspace of codimension 1 of $L^2(\Omega)$, which implies div $(W) = L^2(\Omega)$. Furthermore, div $\left((H^1(\Omega))^2\right) = L^2(\Omega)$.

1.6.3. *Slip condition*

We assume that $\Omega \subset \mathbb{R}^2$ is an open, non-empty, bounded, connected set and its boundary is $\partial\Omega = \overline{\Gamma}_D \cup \overline{\Gamma}_S$, $leng(\Gamma_D) > 0$, $leng(\Gamma_S) > 0$. We impose the condition $\overline{\Gamma}_D \cap \overline{\Gamma}_S = \emptyset$, which is more restrictive than for the case with the Dirichlet condition over one part of the boundary and the Neumann condition over the rest. An example where the two boundaries do not touch each other is $\Omega = \{\mathbf{x} \in \mathbb{R}^2;\ 1 < |\mathbf{x}| < 2\}$ and $\Gamma_D = \{\mathbf{x} \in \mathbb{R}^2;\ |\mathbf{x}| = 1\}$, $\Gamma_S = \{\mathbf{x} \in \mathbb{R}^2;\ |\mathbf{x}| = 2\}$ and $|\mathbf{x}| = \sqrt{x_1^2 + x_2^2}$. Additionally, we want Γ_S to be of class C^3 and Γ_D to be Lipschitz. We use $\mathbf{t}$ to denote the unit vector perpendicular to $\mathbf{n}$ on a point on the boundary. We can choose $\mathbf{t}$ such that the frame $(\mathbf{t}, \mathbf{n})$ is positively oriented, but that is not important.

We are looking for the velocity $\mathbf{v} = (v_1, v_2) : \overline{\Omega} \to \mathbb{R}^2$ and the pressure $p : \overline{\Omega} \to \mathbb{R}$ satisfying [1.43], [1.44], [1.49] and

$$\mathbf{v} \cdot \mathbf{n} = 0, \quad \text{on } \Gamma_S, \tag{1.53}$$

$$(\sigma\mathbf{n}) \cdot \mathbf{t} = 0, \quad \text{on } \Gamma_S. \tag{1.54}$$

Equation [1.53] is called the slip condition. If $(\mathbf{v}, p)$ is a solution, then $(\mathbf{v}, p+const)$ is also a solution, because $\nabla(p+const) = \nabla p$ and $(\sigma\mathbf{n})\cdot\mathbf{t}$ does not depend upon p:

$$0 = (\sigma\mathbf{n}) \cdot \mathbf{t} = (-p\mathbf{n} + 2\mu\epsilon(\mathbf{v})\,\mathbf{n}) \cdot \mathbf{t} = 2\mu\,(\epsilon(\mathbf{v})\,\mathbf{n}) \cdot \mathbf{t}.$$

As for the Neumann condition, after multiplying [1.43] by $\mathbf{w} = 0$ on Γ_D, we obtain the term $\int_{\Gamma_N} (\sigma\mathbf{n})\cdot\mathbf{w}\, ds$. We have the decomposition $\mathbf{w} = (\mathbf{w}\cdot\mathbf{n})\mathbf{n} + (\mathbf{w}\cdot\mathbf{t})\mathbf{t}$ which gives

$$(\sigma\mathbf{n})\cdot\mathbf{w} = (\sigma\mathbf{n})\cdot((\mathbf{w}\cdot\mathbf{n})\mathbf{n} + (\mathbf{w}\cdot\mathbf{t})\mathbf{t}) = (\mathbf{w}\cdot\mathbf{n})(\sigma\mathbf{n})\cdot\mathbf{n} + (\mathbf{w}\cdot\mathbf{t})(\sigma\mathbf{n})\cdot\mathbf{t}.$$

We can introduce the mixed formulation. This time we use the spaces

$$W = \{\mathbf{w} \in (H^1(\Omega))^2;\ \mathbf{w} = 0 \text{ on } \Gamma_D\}, \quad Q = L_0^2(\Omega), \quad M = H^{1/2}(\Gamma_S)$$

and $M' = H^{-1/2}(\Gamma_S)$ is the dual of M. We assume that $\mathbf{f} \in (L^2(\Omega))^2$. Find $\mathbf{v} \in W$, $p \in Q$ and $\eta \in M'$ such that

$$2\mu \int_\Omega \epsilon(\mathbf{v}) : \epsilon(\mathbf{w})\, d\mathbf{x} - \int_\Omega (\nabla\cdot\mathbf{w})\, p\, d\mathbf{x} - \langle \eta, \mathbf{w}\cdot\mathbf{n}\rangle_{1/2,\Gamma_S}$$

$$= \int_\Omega \mathbf{f}\cdot\mathbf{w}\, d\mathbf{x}, \quad \forall \mathbf{w} \in W \qquad [1.55]$$

$$-\int_\Omega (\nabla\cdot\mathbf{v})\, q\, d\mathbf{x} = 0, \quad \forall q \in Q \qquad [1.56]$$

$$-\langle \zeta, \mathbf{v}\cdot\mathbf{n}\rangle_{1/2,\Gamma_S} = 0, \quad \forall \zeta \in M' \qquad [1.57]$$

where $\langle\cdot,\cdot\rangle_{1/2,\Gamma_S}$ is the duality product of $H^{-1/2}(\Gamma_S)$ and $H^{1/2}(\Gamma_S)$. Under the previous hypotheses, if $\mathbf{w} \in W$ then $\mathbf{w}\cdot\mathbf{n} \in M$. If Γ_S is only Lipschitz, the above conclusion is not correct. If the solution of [1.55]–[1.57] is regular, we can interpret η as $(\sigma\mathbf{n})\cdot\mathbf{n}$ and we have [1.54]. The duality product of $H^{-1/2}(\Gamma_S)$ and $H^{1/2}(\Gamma_S)$ is an extension of the scalar product of $L^2(\Gamma_S)$ in the sense that, if $\eta \in L^2(\Gamma_S)$, $\zeta \in H^{1/2}(\Gamma_S)$, then $\langle\eta,\zeta\rangle_{1/2,\Gamma_S} = \int_{\Gamma_S} \eta\tau ds$, [GIR 86, p. 8].

Equation [1.57] represents the *weak treatment* of [1.53]. There also exists an approach, where [1.53] is imposed in the definition of W, which is known as *strong treatment*.

With the notation

$$a : W \times W \to \mathbb{R},\ a(\mathbf{v},\mathbf{w}) = \int_\Omega 2\mu\, \epsilon(\mathbf{v}) : \epsilon(\mathbf{w})\, d\mathbf{x}$$

$$b : W \times (Q \times M') \to \mathbb{R},\ b(\mathbf{w},(q,\zeta)) = -\int_\Omega (\nabla\cdot\mathbf{w})\, q\, d\mathbf{x} - \langle\zeta, \mathbf{w}\cdot\mathbf{n}\rangle_{1/2,\Gamma_S}$$

we have a mixed problem.

Problem [1.55]–[1.57] has a unique solution $\mathbf{v} \in W$, $p \in Q$ and $\eta \in M'$ (see [VER 87], [BÄN 00]).

For the finite element approximation, we can use globally continuous $\mathbb{P}_2$ + *bubble* over each triangle for the velocity, globally continuous $\mathbb{P}_1$ over each triangle for the pressure, and globally discontinuous $\mathbb{P}_0$ over each side of Γ_S for η. Let Ω_h be a polyhedral set that approximates Ω. In the discrete case, instead of $\langle \zeta_h, \mathbf{w}_h \cdot \mathbf{n}_h \rangle_{1/2,\Gamma_S}$, we can put $\int_{\Gamma_S} \zeta_h (\mathbf{w}_h \cdot \mathbf{n}_h) ds$. We will introduce the spaces with finite dimension

$$
\begin{aligned}
W_h = \{\mathbf{w}_h = (w_h^1, w_h^2) \in (C^0(\overline{\Omega}_h))^2;\ w_h = 0 \text{ on } \Gamma_{D,h}, \\
\forall T \in \mathcal{T}_h,\ w^i_{h|T} \in (\mathbb{P}_2 + b)(T),\ i = 1, 2\}
\end{aligned}
$$

$$
Q_h = \{q_h \in C^0(\overline{\Omega}_h);\ \forall T \in \mathcal{T}_h,\ q_{h|T} \in \mathbb{P}_1(T),\ \int_{\Omega_h} q_h d\mathbf{x} = 0\}
$$

$$
M_h = \{\zeta_h : \Gamma_{S,h} \to \mathbb{R};\ \forall S \in \Gamma_{S,h},\ \zeta_{h|S} \in \mathbb{P}_0(S)\}
$$

and the maps $a_h : W_h \times W_h \to \mathbb{R}$, $b_h : W_h \times (Q_h \times M_h) \to \mathbb{R}$

$$
a_h(\mathbf{v}_h, \mathbf{w}_h) = \int_{\Omega_h} 2\mu\, \boldsymbol{\epsilon}(\mathbf{v}_h) : \boldsymbol{\epsilon}(\mathbf{w}_h)\, d\mathbf{x}
$$

$$
b_h(\mathbf{w}_h, (q_h, \zeta_h)) = -\int_{\Omega_h} (\nabla \cdot \mathbf{w}_h)\, q_h\, d\mathbf{x} - \int_{\Gamma_{S,h}} \zeta_h (\mathbf{w}_h \cdot \mathbf{n}_h) ds.
$$

The discrete problem: find $\mathbf{v}_h \in W_h$, $(p_h, \eta_h) \in Q_h \times M_h$

$$
\begin{aligned}
a_h(\mathbf{v}_h, \mathbf{w}_h) + b_h(\mathbf{w}_h, (p_h, \eta_h)) &= \int_{\Omega_h} \mathbf{f}_h \cdot \mathbf{w}_h\, d\mathbf{x}, \quad \forall \mathbf{w}_h \in W_h \\
b_h(\mathbf{v}_h, (q_h, \zeta_h)) &= 0, \quad \forall (q_h, \zeta_h) \in Q_h \times M_h
\end{aligned}
$$

has a unique solution [VER 87], [BÄN 00].

1.7. Numerical tests

For all the tests in the book, we have used `FreeFem++`, [HEC 12].

1.7.1. *Homogeneous Dirichlet condition with exact solution*

We use the example from [BER 79]. The domain is $\Omega =]0,1[\times]0,1[$ and $\mu = 1$. The exact solution is

$$v_1(x_1,x_2) = -256x_2(x_2-1)(2x_2-1)x_1^2(x_1-1)^2,$$

$$v_2(x_1,x_2) = 256x_1(x_1-1)(2x_1-1)x_2^2(x_2-1)^2,$$

$$p(x_1,x_2) = (x_1-0.5)(x_2-0.5).$$

With the notation $\boldsymbol{\sigma} = -p\mathbf{I} + 2\mu\boldsymbol{\epsilon}(\mathbf{v})$, we calculate $\mathbf{f} = -\nabla\cdot(\boldsymbol{\sigma})$. We want to solve the Stokes equations with homogeneous Dirichlet conditions [1.1]–[1.3]. We have used triangulations with the parameter $h = 2^{-j}$ with $j = 4,5,6,7$. First, we used the finite elements $\mathbb{P}_1 + b$ for the velocity and $\mathbb{P}_1$ for the pressure. The errors are presented in Figures 1.1 and 1.2.

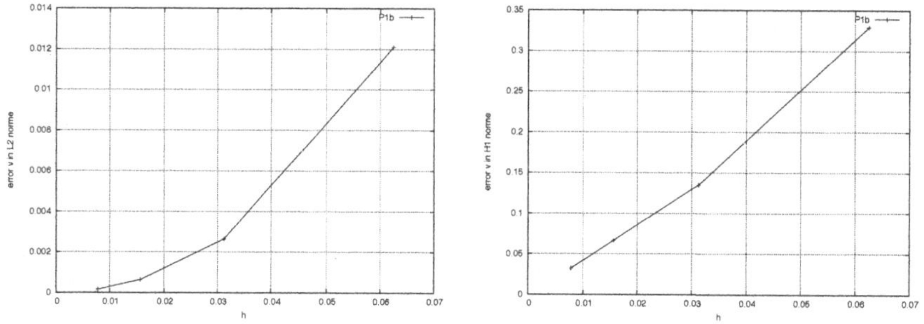

Figure 1.1. *The errors in the velocity with norm $L^2(\Omega)$ (left) and norm $H^1(\Omega)$ (right) for $\mathbb{P}_1 + b/\mathbb{P}_1$*

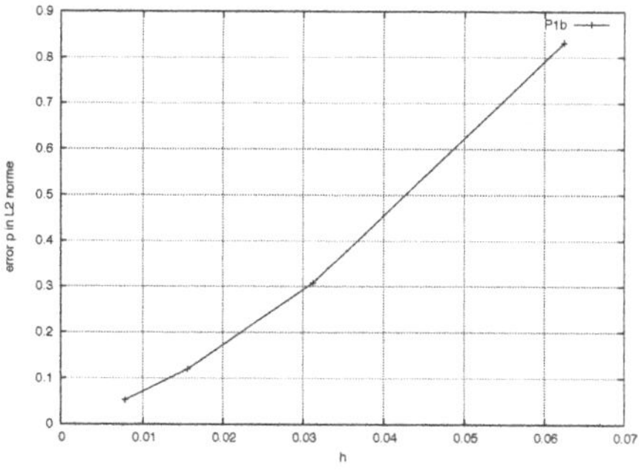

Figure 1.2. *The errors in the pressure with norm de $L^2(\Omega)$ for $\mathbb{P}_1 + b/\mathbb{P}_1$*

We have also repeated the tests with the finite elements $\mathbb{P}_2$ for the velocity and $\mathbb{P}_1$ for the pressure. We observe in Figures 1.3 and 1.4 that the results are better using $\mathbb{P}_2$ for the velocity.

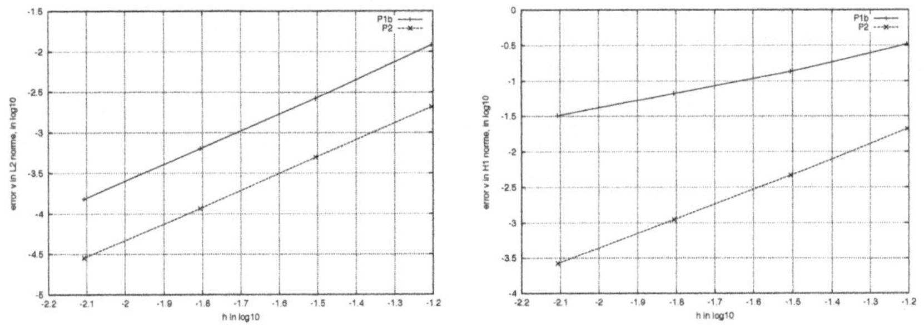

Figure 1.3. *The errors in the velocity with norm $L^2(\Omega)$ (left) and norm $H^1(\Omega)$ (right), logarithmic scale*

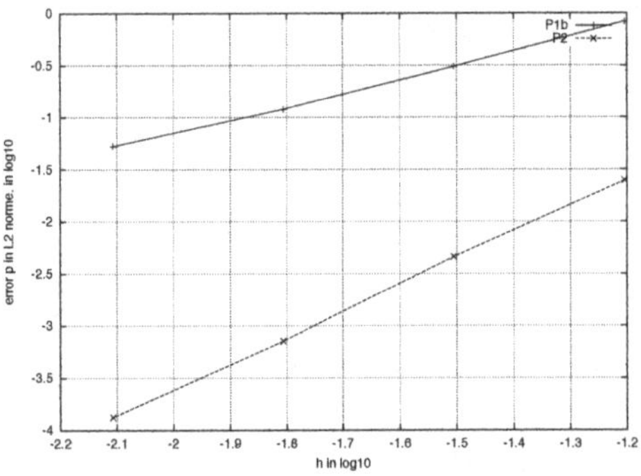

Figure 1.4. *The errors in the pressure with norm $L^2(\Omega)$, logarithmic scale*

In Figures 1.3 and 1.4, we have used a log-log scale (the advantage of this choice is explained in the following). If for different values of h, the points $(\log_{10}(h), \log_{10}(\|\mathbf{v} - \mathbf{v}_h\|_{0,\Omega}))$, for example, are on a straight line of slope q, *i.e.*

$$\log_{10}(\|\mathbf{v} - \mathbf{v}_h\|_{0,\Omega}) = q \log_{10}(h) + c,$$

then $\|\mathbf{v} - \mathbf{v}_h\|_{0,\Omega} = h^q 10^c$ and q is called the order of convergence. In general, the points are not exactly aligned, and to calculate the order, we use the least-squares fitting line. We can use `Scilab` and obtain the following orders of convergence for $\mathbb{P}_1 + b/\mathbb{P}_1$: 2.1 for the velocity in the norm $L^2(\Omega)$, 1.1 for the velocity in the norm $H^1(\Omega)$, 1.3 for the pressure in the norm $L^2(\Omega)$. The results agree with [1.42].

For $\mathbb{P}_2/\mathbb{P}_1$, we have the following orders: 2.06 for the velocity in the norm $L^2(\Omega)$, 2.09 for the velocity in the norm $H^1(\Omega)$, 2.53 for the pressure in the norm $L^2(\Omega)$.

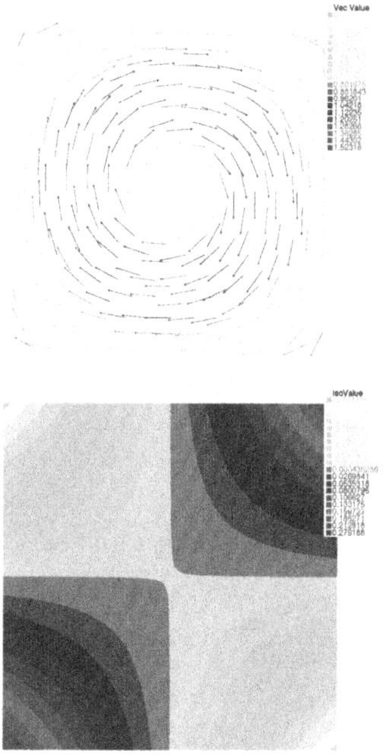

Figure 1.5. *The velocity obtained with $h = 2^{-4}$ and $\mathbb{P}_1 + b$ (above) and the pression obtained with $h = 2^{-7}$ and $\mathbb{P}_2$ (below). For a color version of this figure, see www.iste.co.uk/murea/schemes.zip*

We will now discuss the solution of the linear system obtained after finite element discretization of the Stokes problem. We have seen that for finite-dimension spaces

$$W_h = \{\mathbf{w}_h = (w_h^1, w_h^2) \in (C^0(\overline{\Omega}))^2;\ w_h = 0 \text{ on } \partial\Omega,$$
$$\forall T \in \mathcal{T}_h,\ w^i_{h|T} \in (\mathbb{P}_1 + b)(T),\ i = 1, 2\} \qquad [1.58]$$

$$Q_h = \{q_h \in C^0(\overline{\Omega});\ \forall T \in \mathcal{T}_h,\ q_{h|T} \in \mathbb{P}_1(T),\ \int_\Omega q_h d\mathbf{x} = 0\} \qquad [1.59]$$

the discrete Stokes problem: find $\mathbf{v}_h \in W_h$, $p_h \in Q_h$ such that

$$a(\mathbf{v}_h, \mathbf{w}_h) + b(\mathbf{w}_h, p_h) = \int_{\Omega_h} \mathbf{f} \cdot \mathbf{w}_h \, d\mathbf{x}, \quad \forall \mathbf{w}_h \in W_h \tag{1.60}$$

$$b(\mathbf{v}_h, q_h) = 0, \quad \forall q_h \in Q_h \tag{1.61}$$

has a unique solution. The constraint $\int_\Omega q_h d\mathbf{x} = 0$ in [1.59] makes it difficult to construct a finite element basis for Q_h. We prefer to work with

$$\widetilde{Q}_h = \{\widetilde{q}_h \in C^0(\overline{\Omega});\ \forall T \in \mathcal{T}_h,\ \widetilde{q}_{h|T} \in \mathbb{P}_1(T)\}.$$

The problem is: find $\mathbf{v}_h \in W_h$, $p_h \in \widetilde{Q}_h$ such that

$$a(\mathbf{v}_h, \mathbf{w}_h) + b(\mathbf{w}_h, p_h) = \int_{\Omega_h} \mathbf{f} \cdot \mathbf{w}_h \, d\mathbf{x}, \quad \forall \mathbf{w}_h \in W_h \tag{1.62}$$

$$b(\mathbf{v}_h, \widetilde{q}_h) = 0, \quad \forall \widetilde{q}_h \in \widetilde{Q}_h \tag{1.63}$$

and $\int_\Omega p_h d\mathbf{x} = 0$.

One solution of [1.62]–[1.63], which satisfies $\int_\Omega p_h d\mathbf{x} = 0$ is the solution of [1.60]–[1.61]. If $\mathbf{v}_h \in W_h$, $\widetilde{p}_h \in \widetilde{Q}_h$ is a solution of [1.62]–[1.63], by setting $p_h = \widetilde{p}_h - C$, where $C = \frac{\int_\Omega \widetilde{p}_h d\mathbf{x}}{area(\Omega)}$, then $p_h \in Q_h$. Additionally, $b(\mathbf{w}_h, p_h) = b(\mathbf{w}_h, \widetilde{p}_h) - b(\mathbf{w}_h, C) = 0 - C \int_\Omega (\nabla \cdot \mathbf{w}_h) \, d\mathbf{x} = -C \int_{\partial\Omega} \mathbf{w}_h \cdot \mathbf{n} \, d\mathbf{x} = 0$. In turn, let $\mathbf{v}_h \in W_h$, $p_h \in Q_h$ be the solution of [1.60]–[1.61] and take arbitrary $\widetilde{q}_h \in \widetilde{Q}_h$. We set $C = \frac{\int_\Omega \widetilde{q}_h d\mathbf{x}}{area(\Omega)}$, $q_h = \widetilde{q}_h - C$ and we have $q_h \in Q_h$. Hence, $b(\mathbf{v}_h, \widetilde{q}_h) = b(\mathbf{v}_h, q_h) + b(\mathbf{v}_h, C) = 0 - C \int_\Omega (\nabla \cdot \mathbf{v}_h) \, d\mathbf{x} = -C \int_{\partial\Omega} \mathbf{v}_h \cdot \mathbf{n} \, d\mathbf{x} = 0$, giving [1.63].

The discrete system [1.62]–[1.63] is equivalent to the linear system

$$\begin{pmatrix} A & \widetilde{B}^t \\ \widetilde{B} & 0 \end{pmatrix} \begin{pmatrix} V \\ \widetilde{P} \end{pmatrix} = \begin{pmatrix} F \\ 0 \end{pmatrix} \tag{1.64}$$

where A is a real, symmetric, positive-definite $n \times n$ matrix and $\widetilde{B}$ is a real $m \times n$ matrix, $m < n$ and $rank(\widetilde{B}) < m$. System [1.64] has infinite solutions, but V is uniquely determined. Traditionally, to solve this type of linear system, we would use the Uzawa iteration or the augmented Lagrangian method, [FOR 82]. Now, however, there exist powerful versions of the direct

LU method for sparse matrices which are completely satisfactory. We can also replace [1.62]–[1.63] by: find $\mathbf{v}_h^\epsilon \in W_h$, $p_h^\epsilon \in \widetilde{Q}_h$ such that

$$a(\mathbf{v}_h^\epsilon, \mathbf{w}_h) + b(\mathbf{w}_h, p_h^\epsilon) = \int_{\Omega_h} \mathbf{f} \cdot \mathbf{w}_h \, d\mathbf{x}, \quad \forall \mathbf{w}_h \in W_h \tag{1.65}$$

$$b(\mathbf{v}_h^\epsilon, \widetilde{q}_h) - \epsilon \int_\Omega p_h^\epsilon \widetilde{q}_h \, d\mathbf{x} = 0, \quad \forall \widetilde{q}_h \in \widetilde{Q}_h \tag{1.66}$$

where $\epsilon > 0$ is a small parameter. The associated linear system is

$$\begin{pmatrix} A & \widetilde{B}^t \\ \widetilde{B} & -\epsilon M \end{pmatrix} \begin{pmatrix} V^\epsilon \\ P^\epsilon \end{pmatrix} = \begin{pmatrix} F \\ 0 \end{pmatrix} \tag{1.67}$$

where M is a real, symmetric, positive-definite $m \times m$ matrix. By using the Cholesky factorization $M = R^T R$ where R is an upper triangular matrix, we deduce that $A + \frac{1}{\epsilon}\widetilde{B}^t M^{-1} \widetilde{B}$ is symmetric, positive-definite. Hence, system [1.67] has the unique solution

$$V^\epsilon = \left(A + \frac{1}{\epsilon}\widetilde{B}^t M^{-1} \widetilde{B}\right)^{-1} F, \quad P^\epsilon = \frac{1}{\epsilon} M^{-1} \widetilde{B} V^\epsilon$$

thus, [1.65]–[1.66] has a unique solution. Since the constant function 1 is in $\widetilde{Q}_h$, by putting $\widetilde{q}_h = 1$ in [1.66], we have

$$\epsilon \int_\Omega p_h^\epsilon d\mathbf{x} = -\int_\Omega \nabla \cdot \mathbf{v}_h^\epsilon \, d\mathbf{x} = -\int_{\partial\Omega} \mathbf{v}_h^\epsilon \cdot \mathbf{n} \, d\mathbf{x} = 0,$$

thus, $\int_\Omega p_h^\epsilon d\mathbf{x} = 0$. We can show that

$$\left\|\mathbf{v}_h - \mathbf{v}_h^\epsilon\right\|_{1,\Omega} + \left\|p_h - p_h^\epsilon\right\|_{0,\Omega} \leq \epsilon C.$$

In the numerical tests, we can solve [1.67] with an augmented version of the direct LU method.

For a triangulation of *nt* triangles and *nv* vertices, using the finite elements $\mathbb{P}_1 + b/\mathbb{P}_1$, if we do not take account of the boundary conditions, we obtain $n = 2 \times (nv + nt)$ and $m = nv$; thus, the dimension of the linear system is $n + m$. For $h = 2^{-4}$, we have a triangulation of $nt = 600$ and $nv = 333$, we obtain $n + m = 2199$. If we use $\mathbb{P}_2/\mathbb{P}_1$, then $n + m = 2863$. For $h = 2^{-7}$, $n + m = 137824$ for $\mathbb{P}_1 + b$ and $n + m = 177494$ for $\mathbb{P}_2$.

1.7.2. *Neumann condition*

Let Ω be the quarter of a ring of Figure 1.6 with $R_1 = 5$ and $R_2 = 4$, $\Gamma_N =]AB[\cup]CD[$, $\Gamma_D = \partial\Omega \setminus ([AB] \cup [CD])$. We want to solve the Stokes equations for $\mu = 0.035$, $\mathbf{f} = (0,0)$ with the homogeneous Dirichlet conditions on Γ_D and the traction conditions $\mathbf{h} = (20000, 0)$ on $]AB[$ and $\mathbf{h} = (0,0)$ on $]CD[$.

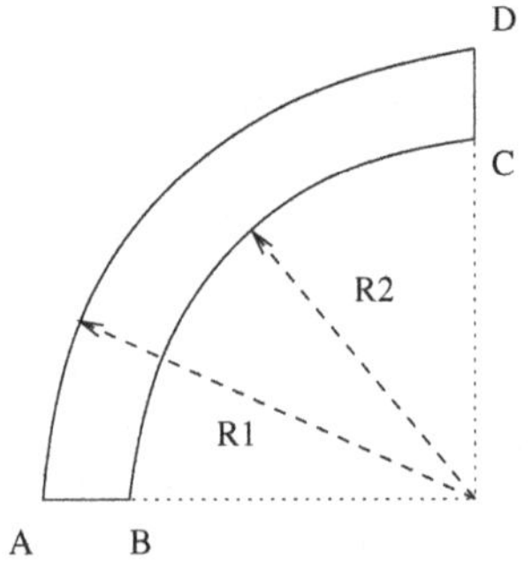

Figure 1.6. *A quarter of a ring*

The numerical results using a triangulation of 4043 triangles and 2160 vertices are presented in Figure 1.7. The pressure is uniquely determined in $L^2(\Omega)$ and $\int_\Omega p_h \, d\mathbf{x} = 70717.8$.

1.7.3. *Slip condition: flow in a quarter disc*

Let Ω be the quarter disc of radius $R = 1$, $\Gamma_D =]AB] \cup]BC[$, Γ_S be the circular arc CA, $\mu = 0.035$, $\mathbf{f} = (0,0)$ (see Figure 1.8). We want to solve the Stokes equations [1.43], [1.44] with the slip conditions [1.53], [1.54] and the non-homogeneous Dirichlet conditions $\mathbf{v} = (-x_2, 0)$ on $]AB]$, $\mathbf{v} = (0, x_1)$ on $]BC[$.

The exact solution is: $\mathbf{v}(x_1, x_2) = (-x_2, x_1)$ and $p(x_1, x_2) = 0$.

For the numerical test, we have used a triangulation of 182 triangles, 110 vertices and the following finite elements: globally continuous $\mathbb{P}_2$ + *bubble* over each triangle for the velocity, globally continuous $\mathbb{P}_1$ over each triangle for the pressure and globally discontinuous $\mathbb{P}_0$ over each side of Γ_S for η, with a weak treatment of the slip condition.

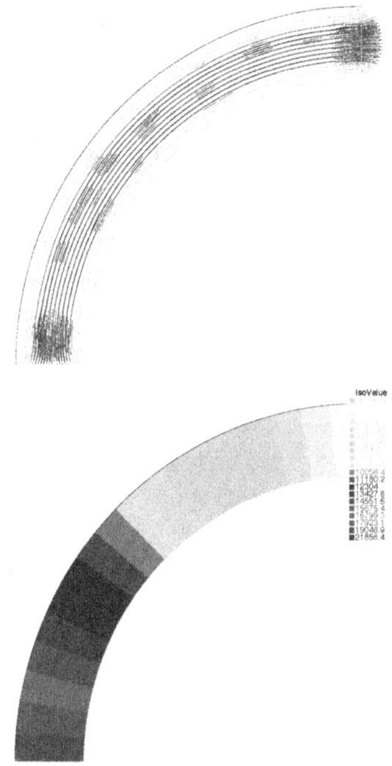

Figure 1.7. *The velocity (above) and the pressure (below) obtained with $\mathbb{P}_1 + b/\mathbb{P}_1$. For a color version of this figure, see www.iste.co.uk/murea/schemes.zip*

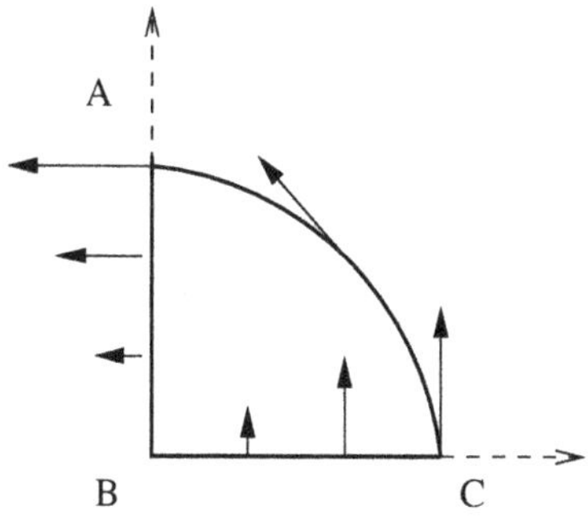

Figure 1.8. *A quarter of a disc*

The velocities are presented in Figure 1.9.

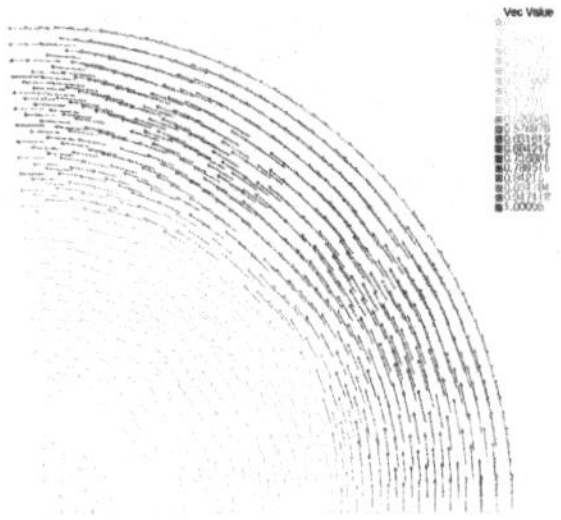

Figure 1.9. *The fluid velocities. For a color version of this figure, see www.iste.co.uk/murea/schemes.zip*

2

Numerical Schemes for the Navier-Stokes Equations

2.1. Navier-Stokes equations

We shall now study the Navier-Stokes equations in a fixed open set $\Omega \subset \mathbb{R}^2$ in the time interval $[0, T]$. We assume that Ω is bounded, connected and its boundary $\partial\Omega$ is Lipschitz.

We are looking for the velocity $\mathbf{v} = (v_1, v_2) : \overline{\Omega} \times [0, T] \rightarrow \mathbb{R}^2$ and the pressure $p : \overline{\Omega} \times [0, T] \rightarrow \mathbb{R}$, such that

$$\rho\left(\frac{\partial \mathbf{v}}{\partial t} + (\mathbf{v} \cdot \nabla)\mathbf{v}\right) - 2\mu\nabla \cdot \boldsymbol{\epsilon}(\mathbf{v}) + \nabla p = \mathbf{f}, \quad \text{in } \Omega\times]0, T[, \qquad [2.1]$$

$$\nabla \cdot \mathbf{v} = 0, \quad \text{in } \Omega\times]0, T[, \qquad [2.2]$$

$$\mathbf{v} = 0, \quad \text{on } \partial\Omega\times]0, T[, \qquad [2.3]$$

$$\mathbf{v}(\mathbf{x}, 0) = \mathbf{v}_0(\mathbf{x}), \quad \forall \mathbf{x} \in \Omega \qquad [2.4]$$

where

– ρ, the mass density, and μ, the dynamic viscosity, are two positive constants,

– $\mathbf{f} = (f_1, f_2) : \Omega\times]0, T[\rightarrow \mathbb{R}^2$ are the body forces, for example, $f_1 = 0$ and $f_2 = -\rho\, 9.81$ for the gravitational forces,

– $\mathbf{v}_0 : \Omega \rightarrow \mathbb{R}^2$ is the initial fluid velocity.

We use the notation

– $\boldsymbol{\epsilon}(\mathbf{v}) = \frac{1}{2}\left(\nabla\mathbf{v} + (\nabla\mathbf{v})^T\right)$ for the (linearized) strain rate tensor,

– $\sigma = -p\mathbf{I} + 2\mu\boldsymbol{\epsilon}(\mathbf{v})$ for the stress tensor, where $\mathbf{I}$ is the unit (or identity) matrix.

The gradients of a scalar function $q : \Omega \to \mathbb{R}$ and of a vector-valued function $\mathbf{w} = (w_1, w_2) : \Omega \to \mathbb{R}^2$ are denoted by the following equation:

$$\nabla q = \begin{pmatrix} \frac{\partial q}{\partial x_1} \\ \frac{\partial q}{\partial x_2} \end{pmatrix}, \quad \nabla \mathbf{w} = \begin{pmatrix} \frac{\partial w_1}{\partial x_1} & \frac{\partial w_1}{\partial x_2} \\ \frac{\partial w_2}{\partial x_1} & \frac{\partial w_2}{\partial x_2} \end{pmatrix}.$$

The divergence operator acting on a vector-valued function $\mathbf{w} = (w_1, w_2) : \Omega \to \mathbb{R}^2$ and on a tensor $\boldsymbol{\sigma} = \left(\sigma_{ij}\right)_{1 \le i,j \le 2}$ is denoted by the following equation:

$$\nabla \cdot \mathbf{w} = \frac{\partial w_1}{\partial x_1} + \frac{\partial w_2}{\partial x_2}, \quad \nabla \cdot \boldsymbol{\sigma} = \begin{pmatrix} \frac{\partial \sigma_{11}}{\partial x_1} + \frac{\partial \sigma_{12}}{\partial x_2} \\ \frac{\partial \sigma_{21}}{\partial x_1} + \frac{\partial \sigma_{22}}{\partial x_2} \end{pmatrix}.$$

We also use

$$(\mathbf{v} \cdot \nabla)\mathbf{w} = \begin{pmatrix} v_1 \frac{\partial w_1}{\partial x_1} + v_2 \frac{\partial w_1}{\partial x_2} \\ v_1 \frac{\partial w_2}{\partial x_1} + v_2 \frac{\partial w_2}{\partial x_2} \end{pmatrix}.$$

If $(X, \|\cdot\|_X)$ is a normed vector space, we use $C([0,T], X)$ or $C^0([0,T], X)$ to denote the set of continuous functions defined over $[0,T]$ with values in X. We can equip the space of functions with the norm

$$\|f\|_{C([0,T],X)} = \sup_{t \in [0,T]} \|f(t)\|_X.$$

We say that the function $f : [0,T] \to X$ is differentiable in $t_0 \in [0,T]$ if there exists an element denoted $f'(t_0)$ or $\frac{df}{dt}(t_0)$ in X, such that

$$\lim_{t \to t_0, t \neq t_0} \frac{f(t) - f(t_0)}{t - t_0} = f'(t_0).$$

We shall use the spaces $L^2(\Omega)$, $H^1(\Omega)$ with the scalar products and norms

$$(v, w)_{0,\Omega} = \int_\Omega v(\mathbf{x}) w(\mathbf{x})\, d\mathbf{x}, \quad \|v\|_{0,\Omega} = \sqrt{(v, v)_{0,\Omega}},$$

and

$$(v, w)_{1,\Omega} = \int_\Omega v(\mathbf{x})w(\mathbf{x})\,d\mathbf{x} + \int_\Omega \nabla v \cdot \nabla w\,d\mathbf{x}, \quad \|v\|_{1,\Omega} = \sqrt{(v, v)_{1,\Omega}}$$

respectively, and we note that

$$L_0^2(\Omega) = \{q \in L^2(\Omega);\ \int_\Omega q\,d\mathbf{x} = 0\}.$$

2.1.1. *Mixed weak formulation*

We assume that $\mathbf{f} \in C^0\left([0, T], (L^2(\Omega))^2\right)$ and $\mathbf{v}_0 \in (H_0^1(\Omega))^2$.

Find

$$\mathbf{v} \in C^1\left(]0, T[, (L^2(\Omega))^2\right) \cap C^0\left([0, T], (H_0^1(\Omega))^2\right)$$

$$p \in C^0\left([0, T], L_0^2(\Omega)\right)$$

such that

$$\rho\frac{d}{dt}\int_\Omega \mathbf{v}\cdot\mathbf{w}\,d\mathbf{x} + \rho\int_\Omega [(\mathbf{v}\cdot\nabla)\mathbf{v}]\cdot\mathbf{w}\,d\mathbf{x} + 2\mu\int_\Omega \boldsymbol{\epsilon}(\mathbf{v}) : \boldsymbol{\epsilon}(\mathbf{w})\,d\mathbf{x}$$

$$-\int_\Omega (\nabla\cdot\mathbf{w})\,p\,d\mathbf{x} = \int_\Omega \mathbf{f}\cdot\mathbf{w}\,d\mathbf{x}, \quad \forall\mathbf{w} \in (H_0^1(\Omega))^2 \qquad [2.5]$$

$$-\int_\Omega (\nabla\cdot\mathbf{v})\,q\,d\mathbf{x} = 0, \quad \forall q \in L_0^2(\Omega) \qquad [2.6]$$

$$\mathbf{v}(0) = \mathbf{v}_0. \qquad [2.7]$$

In [2.6], we can use $q \in L^2(\Omega)$ instead of $q \in L_0^2(\Omega)$. Let $q \in L^2(\Omega)$, then $q - \frac{1}{area(\Omega)}\int_\Omega q\,d\mathbf{x} \in L_0^2(\Omega)$ and in light of [2.6], we have

$$0 = \int_\Omega (\nabla\cdot\mathbf{v})\left(q - \frac{1}{area(\Omega)}\int_\Omega q\,d\mathbf{x}\right)d\mathbf{x}$$

$$= \int_\Omega (\nabla\cdot\mathbf{v})\,q\,d\mathbf{x} - \frac{1}{area(\Omega)}\int_\Omega q\,d\mathbf{x}\int_\Omega (\nabla\cdot\mathbf{v})\,d\mathbf{x}.$$

However, using Green's formula (see theorem A.5) and the fact that $\mathbf{v} \in (H_0^1(\Omega))^2$, we deduce

$$\int_\Omega (\nabla \cdot \mathbf{v})\, d\mathbf{x} = \int_{\partial\Omega} \mathbf{v} \cdot \mathbf{n}\, ds - \int_\Omega \left(v_1 \frac{\partial 1}{\partial x_1} + v_1 \frac{\partial 1}{\partial x_2} \right) d\mathbf{x} = 0$$

where $\mathbf{n}$ is the unit vector, pointing outwards and normal to the boundary. Thus, [2.6] is equivalent to

$$-\int_\Omega (\nabla \cdot \mathbf{v})\, q\, d\mathbf{x} = 0, \quad \forall q \in L^2(\Omega). \qquad [2.8]$$

If we set $q = \nabla \cdot \mathbf{v}$ in [2.6], we have that $\nabla \cdot \mathbf{v} = 0$ in $L^2(\Omega)$. We introduce the space of divergence-free functions

$$V = \{\mathbf{w} \in (H_0^1(\Omega))^2;\ \nabla \cdot \mathbf{w} = 0\}.$$

In order to simplify the notation, we shall set $W = (H_0^1(\Omega))^2$ and $Q = L_0^2(\Omega)$. The advantage of the mixed weak formulation is that we work with test functions in W, not in V.

2.2. Implicit Euler scheme, semi-implicit treatment of the nonlinear term

We introduce

$$a : W \times W \to \mathbb{R}, \quad a(\mathbf{v}, \mathbf{w}) = \int_\Omega 2\mu\, \boldsymbol{\epsilon}(\mathbf{v}) : \boldsymbol{\epsilon}(\mathbf{w})\, d\mathbf{x},$$

$$b : W \times Q \to \mathbb{R}, \quad b(\mathbf{v}, q) = -\int_\Omega (\nabla \cdot \mathbf{v})\, q\, d\mathbf{x},$$

$$c : W \times W \times W \to \mathbb{R}, \quad c(\mathbf{u}, \mathbf{v}, \mathbf{w}) = \int_\Omega [(\mathbf{u} \cdot \nabla)\mathbf{v}] \cdot \mathbf{w}\, d\mathbf{x}.$$

For a and b, we have the properties (see, for example, section 1.3):

$$\exists \alpha > 0,\ \forall \mathbf{w} \in W, \quad \alpha \|\mathbf{w}\|_{1,\Omega}^2 \le a(\mathbf{w}, \mathbf{w}) \qquad [2.9]$$

$$\exists M_a > 0,\ \forall \mathbf{v}, \mathbf{w} \in W, \quad |a(\mathbf{v}, \mathbf{w})| \le M_a \|\mathbf{v}\|_{1,\Omega} \|\mathbf{w}\|_{1,\Omega} \qquad [2.10]$$

$$\exists \beta > 0, \qquad \inf_{q \in Q, q \neq 0} \sup_{w \in W, w \neq 0} \frac{b(\mathbf{w}, q)}{\|\mathbf{w}\|_{1,\Omega} \|q\|_{0,\Omega}} \geq \beta \tag{2.11}$$

$$\exists M_b > 0, \ \forall \mathbf{w} \in W, \ \forall q \in Q, \quad |b(\mathbf{w}, q)| \leq M_b \|\mathbf{w}\|_{1,\Omega} \|q\|_{0,\Omega} . \tag{2.12}$$

For c, we have the properties (see section 2.5):

$$\forall \mathbf{u} \in V, \forall \mathbf{w} \in W, \quad c(\mathbf{u}, \mathbf{w}, \mathbf{w}) = 0 \tag{2.13}$$

$$\exists M_c > 0, \ \forall \mathbf{u}, \mathbf{v}, \mathbf{w} \in W, \quad |c(\mathbf{u}, \mathbf{v}, \mathbf{w})| \leq M_c \|\mathbf{u}\|_{1,\Omega} \|\mathbf{v}\|_{1,\Omega} \|\mathbf{w}\|_{1,\Omega} . \tag{2.14}$$

Using $0 \leq \left\| \frac{1}{\sqrt{\gamma}} v - \sqrt{\gamma} w \right\|_{0,\Omega}^2$ where $\gamma > 0$, we obtain the following inequality:

$$\forall v, w \in L^2(\Omega), \quad (v, w)_{0,\Omega} \leq \frac{1}{2\gamma} \|v\|_{0,\Omega}^2 + \frac{\gamma}{2} \|w\|_{0,\Omega}^2 . \tag{2.15}$$

We divide the interval $[0, T]$ into $N \in \mathbb{N}^*$ intervals of length $\Delta t = T/N$ and we note that $t_n = n\Delta t$ and $\mathbf{f}^n = \mathbf{f}(t_n)$ for $0 \leq n \leq N$. We shall approximate $\mathbf{v}(t_n)$ and $p(t_n)$ by $\mathbf{v}^n$ and p^n.

2.2.1. *Implicit Euler scheme and semi-implicit treatment of the nonlinear term*

Find $\mathbf{v}^{n+1} \in W$ and $p^{n+1} \in Q$, such that

$$\rho \frac{1}{\Delta t} \int_\Omega \mathbf{v}^{n+1} \cdot \mathbf{w} \, d\mathbf{x} + \rho c(\mathbf{v}^n, \mathbf{v}^{n+1}, \mathbf{w}) + a(\mathbf{v}^{n+1}, \mathbf{w}) + b(\mathbf{w}, p^{n+1})$$

$$= \int_\Omega \mathbf{f}^{n+1} \cdot \mathbf{w} \, d\mathbf{x} + \rho \frac{1}{\Delta t} \int_\Omega \mathbf{v}^n \cdot \mathbf{w} \, d\mathbf{x}, \quad \forall \mathbf{w} \in W \tag{2.16}$$

$$b(\mathbf{v}^{n+1}, q) = 0, \quad \forall q \in Q. \tag{2.17}$$

We initialize with $\mathbf{v}^0 = \mathbf{v}_0$ and we assume that $\nabla \cdot \mathbf{v}_0 = 0$.

We remark that at each step in time, we are solving a linear system. This is a weak mixed problem because we approximate the velocity and the pressure by using different finite elements, for example $\mathbb{P}_2$ for the velocity and $\mathbb{P}_1$ for the pressure or $\mathbb{P}_1$ + *bubble* for the velocity and $\mathbb{P}_1$ for the pressure. The theory of Babuska [BAB 73] and Brezzi [BRE 74] tells us that the problem has a unique solution (see Theorem 1.1). In fact, the system [2.16]–[2.17] is a mixed problem with the forms b defined previously and $\mathcal{A} : W \times W \to \mathbb{R}$ defined by the following equation:

$$\mathcal{A}(\mathbf{v}, \mathbf{w}) = \rho \frac{1}{\Delta t} \int_\Omega \mathbf{v} \cdot \mathbf{w} \, d\mathbf{x} + \rho c(\mathbf{v}^n, \mathbf{v}, \mathbf{w}) + a(\mathbf{v}, \mathbf{w}).$$

In order to obtain the continuity of $\mathcal{A}$, we use [2.10] and [2.14]. The ellipticity is obtained using [2.9] and [2.13]:

$$\begin{aligned}\mathcal{A}(\mathbf{w}, \mathbf{w}) &= \rho \frac{1}{\Delta t} \int_\Omega \mathbf{w} \cdot \mathbf{w} \, d\mathbf{x} + \rho c(\mathbf{v}^n, \mathbf{w}, \mathbf{w}) + a(\mathbf{w}, \mathbf{w}) \\ &= \rho \frac{1}{\Delta t} \|\mathbf{w}\|_{0,\Omega}^2 + a(\mathbf{w}, \mathbf{w}) \geq \alpha \|\mathbf{w}\|_{1,\Omega}^2 .\end{aligned}$$

This scheme is of order one in time. To prove this, we assume that

$$\mathbf{v} \in C^2\left(]0, T[, (H_0^1(\Omega))^2\right)$$

and we set

$$\begin{aligned}r = \rho \int_\Omega \frac{\mathbf{v}(t_{n+1}) - \mathbf{v}(t_n)}{\Delta t} \cdot \mathbf{w} \, d\mathbf{x} + \rho c(\mathbf{v}(t_n), \mathbf{v}(t_{n+1}), \mathbf{w}) \\ + a(\mathbf{v}(t_{n+1}), \mathbf{w}) + b(\mathbf{w}, p(t_{n+1})) - \int_\Omega \mathbf{f}(t_{n+1}) \cdot \mathbf{w} \, d\mathbf{x}.\end{aligned}$$

Taking account of [2.5] for $t = t_{n+1}$, we obtain

$$\begin{aligned}r &= \rho \int_\Omega \frac{\mathbf{v}(t_{n+1}) - \mathbf{v}(t_n)}{\Delta t} \cdot \mathbf{w} \, d\mathbf{x} + \rho c(\mathbf{v}(t_n), \mathbf{v}(t_{n+1}), \mathbf{w}) \\ &\quad - \rho \int_\Omega \mathbf{v}'(t_{n+1}) \cdot \mathbf{w} \, d\mathbf{x} - \rho c(\mathbf{v}(t_{n+1}), \mathbf{v}(t_{n+1}), \mathbf{w}) \\ &= \rho \int_\Omega \left(\frac{\mathbf{v}(t_{n+1}) - \mathbf{v}(t_n)}{\Delta t} - \mathbf{v}'(t_{n+1}) \right) \cdot \mathbf{w} \, d\mathbf{x} \\ &\quad + \rho c(\mathbf{v}(t_{n+1}) - \mathbf{v}(t_n), \mathbf{v}(t_{n+1}), \mathbf{w}).\end{aligned}$$

However, using Taylor's formula, we have

$$\mathbf{v}(t_{n+1}) = \mathbf{v}(t_n) + (\Delta t)\mathbf{v}'(t_{n+1}) + \frac{(\Delta t)^2}{2}\mathbf{v}''(\tau), \quad \tau \in]t_n, t_{n+1}[.$$

Thus,

$$\begin{aligned} r &= \Delta t\,\rho \int_\Omega \frac{1}{2}\mathbf{v}''(\tau)\cdot \mathbf{w}\,d\mathbf{x} + \Delta t\,\rho\, c(\mathbf{v}'(t_{n+1}) + \frac{\Delta t}{2}\mathbf{v}''(\tau), \mathbf{v}(t_{n+1}), \mathbf{w}) \\ &= O(\Delta t). \end{aligned}$$

We now present a result on the stability, adapted from [TEM 01b].

THEOREM 2.1.– *There exists a constant C, independent of* Δt*, such that*

$$\rho \left\|\mathbf{v}^{n+1}\right\|_{0,\Omega}^2 + \Delta t\,\alpha \sum_{k=0}^{n} \left\|\mathbf{v}^{k+1}\right\|_{1,\Omega}^2 \le C, \quad 0 \le n \le N-1. \qquad [2.18]$$

DEMONSTRATION 2.1.– If we set $q = \nabla\cdot\mathbf{v}^{n+1} \in Q$ in [2.17], we deduce that $\nabla\cdot\mathbf{v}^{n+1} = 0$ for $n \ge 0$. Furthermore, we have assumed that $\nabla\cdot\mathbf{v}^0 = 0$. We set $\mathbf{w} = \mathbf{v}^{n+1}$ in [2.16] and, bearing in mind that $c(\mathbf{v}^n, \mathbf{v}^{n+1}, \mathbf{v}^{n+1}) = 0$, $n \ge 0$ due to [2.13] and $b(\mathbf{v}^{n+1}, p^{n+1}) = 0$ due to [2.17], we have

$$\rho\frac{1}{\Delta t}\int_\Omega (\mathbf{v}^{n+1} - \mathbf{v}^n)\cdot\mathbf{v}^{n+1}\,d\mathbf{x} + a(\mathbf{v}^{n+1}, \mathbf{v}^{n+1}) = \int_\Omega \mathbf{f}^{n+1}\cdot\mathbf{v}^{n+1}\,d\mathbf{x}.$$

For the first term, we use the identity

$$2\,(\mathbf{v}-\mathbf{w}, \mathbf{v})_{0,\Omega} = \|\mathbf{v}\|_{0,\Omega}^2 - \|\mathbf{w}\|_{0,\Omega}^2 + \|\mathbf{v}-\mathbf{w}\|_{0,\Omega}^2$$

and for the right-hand side term, we use inequality [2.15] with $\gamma = \alpha$, then [2.9], which gives

$$\begin{aligned} \left(\mathbf{f}^{n+1}, \mathbf{v}^{n+1}\right)_{0,\Omega} &\le \frac{1}{2\alpha}\left\|\mathbf{f}^{n+1}\right\|_{0,\Omega}^2 + \frac{\alpha}{2}\left\|\mathbf{v}^{n+1}\right\|_{0,\Omega}^2 \\ &\le \frac{1}{2\alpha}\left\|\mathbf{f}^{n+1}\right\|_{0,\Omega}^2 + \frac{\alpha}{2}\left\|\mathbf{v}^{n+1}\right\|_{1,\Omega}^2 \le \frac{1}{2\alpha}\left\|\mathbf{f}^{n+1}\right\|_{0,\Omega}^2 + \frac{1}{2}a(\mathbf{v}^{n+1}, \mathbf{v}^{n+1}). \end{aligned}$$

By multiplying by $2\Delta t$, we thus have

$$\rho\left(\left\|\mathbf{v}^{n+1}\right\|_{0,\Omega}^2 - \left\|\mathbf{v}^{n}\right\|_{0,\Omega}^2 + \left\|\mathbf{v}^{n+1} - \mathbf{v}^{n}\right\|_{0,\Omega}^2\right) + 2\Delta t\, a(\mathbf{v}^{n+1}, \mathbf{v}^{n+1})$$
$$\leq \frac{\Delta t}{\alpha}\left\|\mathbf{f}^{n+1}\right\|_{0,\Omega}^2 + \Delta t\, a(\mathbf{v}^{n+1}, \mathbf{v}^{n+1})$$

or, equivalently

$$\rho\left(\left\|\mathbf{v}^{n+1}\right\|_{0,\Omega}^2 - \left\|\mathbf{v}^{n}\right\|_{0,\Omega}^2 + \left\|\mathbf{v}^{n+1} - \mathbf{v}^{n}\right\|_{0,\Omega}^2\right) + \Delta t\, a(\mathbf{v}^{n+1}, \mathbf{v}^{n+1})$$
$$\leq \frac{\Delta t}{\alpha}\left\|\mathbf{f}^{n+1}\right\|_{0,\Omega}^2. \qquad [2.19]$$

We write inequality [2.19] for $n = 0, 1, \ldots, n$ and by summing, we obtain the following inequality:

$$\rho\left(\left\|\mathbf{v}^{n+1}\right\|_{0,\Omega}^2 - \left\|\mathbf{v}^{0}\right\|_{0,\Omega}^2\right) + \Delta t \sum_{k=0}^{n} a(\mathbf{v}^{k+1}, \mathbf{v}^{k+1}) \leq \frac{\Delta t}{\alpha}\sum_{k=0}^{n}\left\|\mathbf{f}^{k+1}\right\|_{0,\Omega}^2$$
$$\leq \frac{(n+1)\Delta t}{\alpha}\max_{t\in[0,T]}\|\mathbf{f}(t)\|_{0,\Omega}^2 \leq \frac{T}{\alpha}\max_{t\in[0,T]}\|\mathbf{f}(t)\|_{0,\Omega}^2$$

and if we use [2.9], we obtain the following inequality:

$$\rho\left\|\mathbf{v}^{n+1}\right\|_{0,\Omega}^2 + \Delta t\,\alpha\sum_{k=0}^{n}\left\|\mathbf{v}^{k+1}\right\|_{1,\Omega}^2 \leq \rho\left\|\mathbf{v}^{0}\right\|_{0,\Omega}^2 + \frac{T}{\alpha}\max_{t\in[0,T]}\|\mathbf{f}(t)\|_{0,\Omega}^2 = C.$$

□

2.2.2. *And the pressure?*

Starting from [2.16], we have

$$b(\mathbf{w}, p^{n+1}) = \int_\Omega \mathbf{f}^{n+1}\cdot\mathbf{w}\,d\mathbf{x} - \rho\int_\Omega \frac{\mathbf{v}^{n+1} - \mathbf{v}^{n}}{\Delta t}\cdot\mathbf{w}\,d\mathbf{x}$$
$$-\rho c(\mathbf{v}^{n}, \mathbf{v}^{n+1}, \mathbf{w}) - a(\mathbf{v}^{n+1}, \mathbf{w}), \quad \forall \mathbf{w}\in W$$

and by using the Cauchy-Schwarz [A.1] inequality, [2.10] and [2.14], we obtain the following inequality:

$$b(\mathbf{w}, p^{n+1}) \leq \left\|\mathbf{f}^{n+1}\right\|_{0,\Omega} \|\mathbf{w}\|_{0,\Omega} + \rho \left\| \frac{\mathbf{v}^{n+1} - \mathbf{v}^n}{\Delta t} \right\|_{0,\Omega} \|\mathbf{w}\|_{0,\Omega}$$

$$+\rho M_c \left\|\mathbf{v}^n\right\|_{1,\Omega} \left\|\mathbf{v}^{n+1}\right\|_{1,\Omega} \|\mathbf{w}\|_{1,\Omega} + M_a \left\|\mathbf{v}^{n+1}\right\|_{1,\Omega} \|\mathbf{w}\|_{1,\Omega}, \quad \forall \mathbf{w} \in W.$$

Thus,

$$\sup_{w \in W, w \neq 0} \frac{b\left(\mathbf{w}, p^{n+1}\right)}{\|\mathbf{w}\|_{1,\Omega}} \leq \left\|\mathbf{f}^{n+1}\right\|_{0,\Omega} + \rho \left\| \frac{\mathbf{v}^{n+1} - \mathbf{v}^n}{\Delta t} \right\|_{0,\Omega}$$

$$+\rho M_c \left\|\mathbf{v}^n\right\|_{1,\Omega} \left\|\mathbf{v}^{n+1}\right\|_{1,\Omega} + M_a \left\|\mathbf{v}^{n+1}\right\|_{1,\Omega}$$

and taking account of [2.11], we have

$$\beta \left\|p^{n+1}\right\|_{0,\Omega} \leq \left\|\mathbf{f}^{n+1}\right\|_{0,\Omega} + \rho \left\| \frac{\mathbf{v}^{n+1} - \mathbf{v}^n}{\Delta t} \right\|_{0,\Omega}$$

$$+\rho M_c \left\|\mathbf{v}^n\right\|_{1,\Omega} \left\|\mathbf{v}^{n+1}\right\|_{1,\Omega} + M_a \left\|\mathbf{v}^{n+1}\right\|_{1,\Omega}$$

and we cannot go further because inequality [2.18] only tells us that $\left\|\mathbf{v}^k\right\|_{0,\Omega}$ and $\Delta t \left\|\mathbf{v}^k\right\|_{1,\Omega}^2$ are bounded.

2.2.3. *Spatial discretization by finite elements*

We can construct vector spaces of finite dimension $W_h \subset W$ and $Q_h \subset Q$ using finite elements, for example $\mathbb{P}_2$ or $\mathbb{P}_1 +$ *bubble* for the velocity and $\mathbb{P}_1$ for the pressure.

The demonstration of theorem 2.1 uses the fact that $c(\mathbf{v}^n, \mathbf{v}^{n+1}, \mathbf{v}^{n+1}) = 0$, due to [2.13]. In the continuous case, equation [2.17] implies that $\nabla \cdot \mathbf{v}^{n+1} = 0$, but the discrete version $\mathbf{v}_h^{n+1} \in W_h$ and

$$b(\mathbf{v}_h^{n+1}, q_h) = 0, \quad \forall q_h \in Q_h$$

no longer implies that $\nabla \cdot \mathbf{v}_h^{n+1} = 0$. In order to obtain a result for the stability for $\mathbf{v}_h^{n+1}$ (see [TEM 01b] for example) we replace the trilinear form c with $\tilde{c}$ defined by $\tilde{c} : W \times W \times W \to \mathbb{R}$

$$\tilde{c}(\mathbf{u}, \mathbf{v}, \mathbf{w}) \overset{déf}{=} \frac{1}{2} \int_\Omega [(\mathbf{u} \cdot \nabla)\mathbf{v}] \cdot \mathbf{w}\, d\mathbf{x} - \frac{1}{2} \int_\Omega [(\mathbf{u} \cdot \nabla)\mathbf{w}] \cdot \mathbf{v}\, d\mathbf{x}.$$

We observe that

$$\tilde{c}(\mathbf{u}, \mathbf{v}, \mathbf{w}) = \frac{1}{2} c(\mathbf{u}, \mathbf{v}, \mathbf{w}) - \frac{1}{2} c(\mathbf{u}, \mathbf{w}, \mathbf{v}).$$

Thus,

$$\tilde{c}(\mathbf{u}, \mathbf{w}, \mathbf{w}) = 0, \quad \forall \mathbf{u}, \mathbf{w} \in W \qquad [2.20]$$

even if $\nabla \cdot \mathbf{u} \neq 0$.

Thanks to Lemma 2.4, we have

$$\tilde{c}(\mathbf{u}, \mathbf{v}, \mathbf{w}) = c(\mathbf{u}, \mathbf{v}, \mathbf{w}), \quad \forall \mathbf{u} \in V, \forall \mathbf{v}, \mathbf{w} \in W.$$

The temporally and spatially discrete scheme is as follows: find $\mathbf{v}_h^{n+1} \in W_h$ and $p_h^{n+1} \in Q_h$, such that

$$\rho \frac{1}{\Delta t} \int_\Omega \mathbf{v}_h^{n+1} \cdot \mathbf{w}\, d\mathbf{x} + \rho \tilde{c}(\mathbf{v}_h^n, \mathbf{v}_h^{n+1}, \mathbf{w}) + a(\mathbf{v}_h^{n+1}, \mathbf{w}) + b(\mathbf{w}, p_h^{n+1})$$

$$= \int_\Omega \mathbf{f}_h^{n+1} \cdot \mathbf{w}\, d\mathbf{x} + \rho \frac{1}{\Delta t} \int_\Omega \mathbf{v}_h^n \cdot \mathbf{w}\, d\mathbf{x}, \quad \forall \mathbf{w}_h \in W_h \qquad [2.21]$$

$$b(\mathbf{v}_h^{n+1}, q_h) = 0, \quad \forall q_h \in Q_h. \qquad [2.22]$$

By using the fact that $W_h \subset W$ and $Q_h \subset Q$, we obtain the discrete versions of [2.9], [2.10], [2.12], [2.14] automatically. For certain pairs of finite elements, such as $\mathbb{P}_2/\mathbb{P}_1$ or $\mathbb{P}_1$ + *bubble*/$\mathbb{P}_1$, the discrete *inf-sup* condition holds and as a result, we establish the existence and uniqueness of [2.21], [2.22].

By using $\tilde{c}(\mathbf{v}_h^n, \mathbf{v}_h^{n+1}, \mathbf{v}_h^{n+1}) = 0$, we obtain an identical inequality to [2.18].

2.3. Implicit Euler scheme and implicit treatment of the nonlinear term

If we replace the term $c(\mathbf{v}^n, \mathbf{v}^{n+1}, \mathbf{w})$ in [2.16] with $c(\mathbf{v}^{n+1}, \mathbf{v}^{n+1}, \mathbf{w})$, *i.e.* the nonlinear term is *treated implicitly* and then at each step in time, we must solve a nonlinear problem. Theorem 2.1 also remains valid in this case and the demonstration is identical.

Find $\mathbf{v}^{n+1} \in W$ and $p^{n+1} \in Q$, such that

$$\rho\frac{1}{\Delta t}\int_\Omega \mathbf{v}^{n+1}\cdot\mathbf{w}\,d\mathbf{x} + \rho c(\mathbf{v}^{n+1}, \mathbf{v}^{n+1}, \mathbf{w}) + a(\mathbf{v}^{n+1}, \mathbf{w}) + b(\mathbf{w}, p^{n+1})$$

$$= \int_\Omega \mathbf{f}^{n+1}\cdot\mathbf{w}\,d\mathbf{x} + \rho\frac{1}{\Delta t}\int_\Omega \mathbf{v}^n\cdot\mathbf{w}\,d\mathbf{x}, \quad \forall\mathbf{w}\in W \qquad [2.23]$$

$$b(\mathbf{v}^{n+1}, q) = 0, \quad \forall q \in Q. \qquad [2.24]$$

We initialize with $\mathbf{v}^0 = \mathbf{v}_0$. In contrast with the semi-implicit scheme, we no longer require $\nabla\cdot\mathbf{v}_0 = 0$ to demonstrate the stability. The problem [2.23]–[2.24] has at least one solution (see [TEM 01b], Theorem 1.2, p. 110). We can obtain the uniqueness under the assumption that $\|\mathbf{f}^{n+1}\|_{0,\Omega}$ and $\|\frac{1}{\Delta t}\mathbf{v}^n\|_{0,\Omega}$ are small (see [TEM 01b], Theorem 1.3, p. 112).

This scheme is of order one in time. To prove this, we assume that

$$\mathbf{v}\in C^2\left(]0,T[,(H_0^1(\Omega))^2\right)$$

and we set

$$r = \rho\int_\Omega \frac{\mathbf{v}(t_{n+1})-\mathbf{v}(t_n)}{\Delta t}\cdot\mathbf{w}\,d\mathbf{x} + \rho c(\mathbf{v}(t_{n+1}), \mathbf{v}(t_{n+1}), \mathbf{w})$$

$$+a(\mathbf{v}(t_{n+1}), \mathbf{w}) + b(\mathbf{w}, p(t_{n+1})) - \int_\Omega \mathbf{f}(t_{n+1})\cdot\mathbf{w}\,d\mathbf{x}.$$

Taking account of [2.5] for $t = t_{n+1}$, we obtain the following equation:

$$r = \rho\int_\Omega \frac{\mathbf{v}(t_{n+1})-\mathbf{v}(t_n)}{\Delta t}\cdot\mathbf{w}\,d\mathbf{x} - \rho\int_\Omega \mathbf{v}'(t_{n+1})\cdot\mathbf{w}\,d\mathbf{x}.$$

However, using Taylor's formula, we have

$$\mathbf{v}(t_{n+1}) = \mathbf{v}(t_n) + (\Delta t)\mathbf{v}'(t_{n+1}) + \frac{(\Delta t)^2}{2}\mathbf{v}''(\tau), \quad \tau \in]t_n, t_{n+1}[.$$

Thus,

$$r = \Delta t \rho \int_\Omega \frac{1}{2}\mathbf{v}''(\tau) \cdot \mathbf{w}\, d\mathbf{x} = O(\Delta t).$$

To solve the nonlinear system, we shall use *Newton's method.*

– *Step 0.* Initialization: $k = 0$, $\mathbf{v}^{n+1,0} = \mathbf{v}^n$, $p^{n+1,0} = p^n$.

– *Step 1.* Find $\delta\mathbf{v}^k \in W$ and $\delta p^k \in Q$ such that

$$\rho\frac{1}{\Delta t}\int_\Omega \delta\mathbf{v}^k \cdot \mathbf{w}\, d\mathbf{x} + \rho c(\delta\mathbf{v}^k, \mathbf{v}^{n+1,k}, \mathbf{w}) + \rho c(\mathbf{v}^{n+1,k}, \delta\mathbf{v}^k, \mathbf{w}) + a(\delta\mathbf{v}^k, \mathbf{w})$$
$$+ b(\mathbf{w}, \delta p^k) + \rho\frac{1}{\Delta t}\int_\Omega \mathbf{v}^{n+1,k} \cdot \mathbf{w}\, d\mathbf{x} + \rho c(\mathbf{v}^{n+1,k}, \mathbf{v}^{n+1,k}, \mathbf{w}) + a(\mathbf{v}^{n+1,k}, \mathbf{w})$$
$$+ b(\mathbf{w}, p^{n+1,k}) - \int_\Omega \mathbf{f}^{n+1} \cdot \mathbf{w}\, d\mathbf{x} - \rho\frac{1}{\Delta t}\int_\Omega \mathbf{v}^n \cdot \mathbf{w}\, d\mathbf{x} = 0, \ \forall \mathbf{w} \in W \quad [2.25]$$

$$b(\delta\mathbf{v}^k, q) = 0, \quad \forall q \in Q \quad [2.26]$$

– *Step 2.* If $\left\|\delta\mathbf{v}^k\right\|_{0,\Omega} < tol$ then STOP.

– *Step 3.* We set $\mathbf{v}^{n+1,k+1} = \mathbf{v}^{n+1,k} + \delta\mathbf{v}^k$, $p^{n+1,k+1} = p^{n+1,k} + \delta p^k$, $k \leftarrow k+1$, go to *Step 1.*

In the case of non-homogeneous Dirichlet boundary conditions, we shall initialize Newton's method with the solution of a Stokes problem.

2.4. Implicit Euler scheme and explicit treatment of the nonlinear term

We assume that Ω is a polygonal set in $\mathbb{R}^2$, open, connected and bounded. Let $W_h \subset W = (H_0^1(\Omega))^2$ and $Q_h \subset Q = L_0^2(\Omega)$ be vector spaces of finite dimensions. Let $\mathbf{f}_h^n \in (L^2(\Omega))^2$ be an approximation of $\mathbf{f}^n \in (L^2(\Omega))^2$.

Find $\mathbf{v}_h^{n+1} \in W_h$ and $p_h^{n+1} \in Q_h$, such that

$$\rho \frac{1}{\Delta t} \int_\Omega \mathbf{v}_h^{n+1} \cdot \mathbf{w}_h \, d\mathbf{x} + a(\mathbf{v}_h^{n+1}, \mathbf{w}_h) + b(\mathbf{w}_h, p_h^{n+1}) = \int_\Omega \mathbf{f}_h^{n+1} \cdot \mathbf{w}_h \, d\mathbf{x}$$

$$+\rho \frac{1}{\Delta t} \int_\Omega \mathbf{v}_h^n \cdot \mathbf{w}_h \, d\mathbf{x} - \rho \tilde{c}(\mathbf{v}_h^n, \mathbf{v}_h^n, \mathbf{w}_h), \quad \forall \mathbf{w}_h \in W_h \qquad [2.27]$$

$$b(\mathbf{v}_h^{n+1}, q_h) = 0, \quad \forall q_h \in Q_h. \qquad [2.28]$$

We initialize with $\mathbf{v}_h^0 \in W_h$.

We shall use the approach presented in [TEM 01b], section 5.3, p. 230. The Poincaré inequality is written as follows:

$$\exists C_P > 0, \ \forall \mathbf{w} \in W, \ \|\mathbf{w}\|_{0,\Omega} \leq C_P \|\nabla \mathbf{w}\|_{0,\Omega}. \qquad [2.29]$$

Since W_h is a vector space of finite dimensions, the norms $\mathbf{w}_h \to \|\mathbf{w}_h\|_{0,\Omega}$ and $\mathbf{w}_h \to \|\nabla \mathbf{w}_h\|_{0,\Omega}$ are equivalent and there exists $S(h) > 0$, such that

$$\forall \mathbf{w}_h \in W_h, \quad \|\nabla \mathbf{w}_h\|_{0,\Omega} \leq S(h) \|\mathbf{w}_h\|_{0,\Omega}. \qquad [2.30]$$

We can show that $\lim_{h \to 0_+} S(h) = \infty$. This inequality has no equivalent in the continuous case $\mathbf{w} \in W$ and for this reason, we work with the completely discretized scheme, [2.27]–[2.28]. In the previous sections, for the semi-implicit and implicit schemes, we worked with spatially continuous versions.

We set $S_1(h) = \sqrt{2} S(h)$. By using Lemma 2.2 and inequality [2.30], for $\mathbf{u}_h, \mathbf{w}_h \in W_h$, we obtain the following equation:

$$|c(\mathbf{u}_h, \mathbf{u}_h, \mathbf{w}_h)| \leq \sqrt{2} \|\mathbf{u}_h\|_{0,\Omega}^{1/2} \|\nabla \mathbf{u}_h\|_{0,\Omega}^{1/2} \|\nabla \mathbf{u}_h\|_{0,\Omega} \|\mathbf{w}_h\|_{0,\Omega}^{1/2} \|\nabla \mathbf{w}_h\|_{0,\Omega}^{1/2}$$

$$\leq \sqrt{2} \|\mathbf{u}_h\|_{0,\Omega}^{1/2} S^{1/2}(h) \|\mathbf{u}_h\|_{0,\Omega}^{1/2} \|\nabla \mathbf{u}_h\|_{0,\Omega} \|\mathbf{w}_h\|_{0,\Omega}^{1/2} S^{1/2}(h) \|\mathbf{w}_h\|_{0,\Omega}^{1/2}$$

$$= \sqrt{2} S(h) \|\mathbf{u}_h\|_{0,\Omega} \|\nabla \mathbf{u}_h\|_{0,\Omega} \|\mathbf{w}_h\|_{0,\Omega}$$

and

$$
\begin{aligned}
|c(\mathbf{u}_h, \mathbf{w}_h, \mathbf{u}_h)| &\leq \sqrt{2}\, \|\mathbf{u}_h\|_{0,\Omega}^{1/2} \|\nabla \mathbf{u}_h\|_{0,\Omega}^{1/2} \|\nabla \mathbf{w}_h\|_{0,\Omega} \|\mathbf{u}_h\|_{0,\Omega}^{1/2} \|\nabla \mathbf{u}_h\|_{0,\Omega}^{1/2} \\
&= \sqrt{2}\, \|\mathbf{u}_h\|_{0,\Omega} \|\nabla \mathbf{u}_h\|_{0,\Omega} \|\nabla \mathbf{w}_h\|_{0,\Omega} \\
&\leq \sqrt{2} S(h)\, \|\mathbf{u}_h\|_{0,\Omega} \|\nabla \mathbf{u}_h\|_{0,\Omega} \|\mathbf{w}_h\|_{0,\Omega}
\end{aligned}
$$

which give

$$
\begin{aligned}
|\tilde{c}(\mathbf{u}_h, \mathbf{u}_h, \mathbf{w}_h)| &= \frac{1}{2} |c(\mathbf{u}_h, \mathbf{u}_h, \mathbf{w}_h) - c(\mathbf{u}_h, \mathbf{w}_h, \mathbf{u}_h)| \\
&\leq \sqrt{2} S(h)\, \|\mathbf{u}_h\|_{0,\Omega} \|\nabla \mathbf{u}_h\|_{0,\Omega} \|\mathbf{w}_h\|_{0,\Omega} \\
&= S_1(h)\, \|\mathbf{u}_h\|_{0,\Omega} \|\nabla \mathbf{u}_h\|_{0,\Omega} \|\mathbf{w}_h\|_{0,\Omega} \, . \qquad [2.31]
\end{aligned}
$$

When we work with homogeneous Dirichlet conditions along the whole boundary, we have the simplified version of Korn's inequality

$$
\forall \mathbf{v} \in (H_0^1(\Omega))^2, \quad \int_\Omega \nabla \mathbf{v} : \nabla \mathbf{v} d\mathbf{x} \leq 2 \int_\Omega \epsilon(\mathbf{v}) : \epsilon(\mathbf{v}) d\mathbf{x}. \qquad [2.32]
$$

We shall see the demonstration later (see proposition 6.2). We set $\nu = \frac{\mu}{\rho}$.

THEOREM 2.2.– *We assume that*

$$
\exists C_0 > 0,\ \forall h > 0, \quad \left\|\mathbf{v}_h^0\right\|_{1,\Omega} \leq C_0 \qquad [2.33]
$$

$$
\exists M > 0,\ \forall h > 0,\ \forall n \in \mathbb{N},\ n\Delta t \leq T, \quad \left\|\mathbf{f}_h^n\right\|_{0,\Omega} \leq M. \qquad [2.34]
$$

We set $K = \max\{\frac{T C_P^2 M^2}{\nu \rho^2} + C_0^2 + \frac{T \nu C_0^4}{2}, 1\}$. *If*

$$
\Delta t\, S_1^2(h) \leq \frac{\nu}{4K} \qquad [2.35]
$$

then,

$$\left\|\mathbf{v}_h^{n+1}\right\|_{0,\Omega}^2 + \frac{1}{2}\sum_{k=0}^{n}\left\|\mathbf{v}_h^{k+1} - \mathbf{v}_h^k\right\|_{0,\Omega}^2 + \Delta t\, \nu \left\|\nabla \mathbf{v}_h^{n+1}\right\|_{0,\Omega}^2$$

$$+\Delta t \frac{\nu}{2}\sum_{k=1}^{n+1}\left\|\nabla \mathbf{v}_h^k\right\|_{0,\Omega}^2 \leq K, \quad 0 \leq n \leq N-1. \qquad [2.36]$$

DEMONSTRATION 2.2.– We set $\mathbf{w}_h = \mathbf{v}_h^{n+1}$ in [2.27] and taking account of [2.28], we have

$$\frac{\rho}{\Delta t}\int_\Omega \mathbf{v}_h^{n+1}\cdot(\mathbf{v}_h^{n+1} - \mathbf{v}_h^n)\, d\mathbf{x} + a(\mathbf{v}_h^{n+1}, \mathbf{v}_h^{n+1}) = \int_\Omega \mathbf{f}_h^{n+1}\cdot \mathbf{v}_h^{n+1}\, d\mathbf{x}$$

$$-\rho\tilde{c}(\mathbf{v}_h^n, \mathbf{v}_h^n, \mathbf{v}_h^{n+1}).$$

We multiply by $\frac{2\Delta t}{\rho}$ and using the identity

$$2\,(\mathbf{v}_h - \mathbf{w}_h, \mathbf{v}_h)_{0,\Omega} = \|\mathbf{v}_h\|_{0,\Omega}^2 - \|\mathbf{w}_h\|_{0,\Omega}^2 + \|\mathbf{v}_h - \mathbf{w}_h\|_{0,\Omega}^2$$

we obtain the following equation:

$$\left\|\mathbf{v}_h^{n+1}\right\|_{0,\Omega}^2 - \left\|\mathbf{v}_h^n\right\|_{0,\Omega}^2 + \left\|\mathbf{v}_h^{n+1} - \mathbf{v}_h^n\right\|_{0,\Omega}^2 + \frac{2\Delta t}{\rho} a(\mathbf{v}_h^{n+1}, \mathbf{v}_h^{n+1})$$

$$= \frac{2\Delta t}{\rho}\int_\Omega \mathbf{f}_h^{n+1}\cdot \mathbf{v}_h^{n+1}\, d\mathbf{x} - 2\Delta t\, \tilde{c}(\mathbf{v}_h^n, \mathbf{v}_h^n, \mathbf{v}_h^{n+1}).$$

Using [2.32], we have

$$\mu\left\|\nabla \mathbf{v}_h^{n+1}\right\|_{0,\Omega}^2 \leq a(\mathbf{v}_h^{n+1}, \mathbf{v}_h^{n+1}).$$

Inequalities [2.15], [2.29] and hypothesis [2.34] give

$$\frac{2}{\rho}\int_\Omega \mathbf{f}_h^{n+1}\cdot\mathbf{v}_h^{n+1}\,d\mathbf{x} = 2\int_\Omega \frac{1}{\rho}\mathbf{f}_h^{n+1}\cdot\mathbf{v}_h^{n+1}\,d\mathbf{x}$$
$$\leq \frac{C_P^2}{\nu}\left\|\frac{1}{\rho}\mathbf{f}_h^{n+1}\right\|_{0,\Omega}^2 + \frac{\nu}{C_P^2}\left\|\mathbf{v}_h^{n+1}\right\|_{0,\Omega}^2 \leq \frac{C_P^2}{\nu}\left\|\frac{1}{\rho}\mathbf{f}_h^{n+1}\right\|_{0,\Omega}^2 + \nu\left\|\nabla\mathbf{v}_h^{n+1}\right\|_{0,\Omega}^2$$
$$\leq \frac{C_P^2}{\nu\rho^2}M^2 + \nu\left\|\nabla\mathbf{v}_h^{n+1}\right\|_{0,\Omega}^2.$$

We deduce

$$\left\|\mathbf{v}_h^{n+1}\right\|_{0,\Omega}^2 - \left\|\mathbf{v}_h^{n}\right\|_{0,\Omega}^2 + \left\|\mathbf{v}_h^{n+1}-\mathbf{v}_h^{n}\right\|_{0,\Omega}^2 + 2\Delta t\,\nu\left\|\nabla\mathbf{v}_h^{n+1}\right\|_{0,\Omega}^2$$
$$\leq \Delta t\frac{C_P^2}{\nu\rho^2}M^2 + \Delta t\,\nu\left\|\nabla\mathbf{v}_h^{n+1}\right\|_{0,\Omega}^2 - 2\Delta t\,\tilde{c}(\mathbf{v}_h^n,\mathbf{v}_h^n,\mathbf{v}_h^{n+1}).$$

Until now, we have followed the same steps as for the demonstration of the stability of the scheme with semi-implicit treatment of the nonlinear term. We shall focus our attention on the term with $\tilde{c}$. We use [2.20] and [2.31] and we obtain the following equation:

$$-2\Delta t\,\tilde{c}(\mathbf{v}_h^n,\mathbf{v}_h^n,\mathbf{v}_h^{n+1}) = -2\Delta t\,\tilde{c}(\mathbf{v}_h^n,\mathbf{v}_h^n,\mathbf{v}_h^{n+1}-\mathbf{v}_h^n)$$
$$\leq 2\Delta t\left|\tilde{c}(\mathbf{v}_h^n,\mathbf{v}_h^n,\mathbf{v}_h^{n+1}-\mathbf{v}_h^n)\right|$$
$$\leq 2\Delta t\,S_1(h)\left\|\mathbf{v}_h^n\right\|_{0,\Omega}\left\|\nabla\mathbf{v}_h^n\right\|_{0,\Omega}\left\|\mathbf{v}_h^{n+1}-\mathbf{v}_h^n\right\|_{0,\Omega}$$

next we use [2.15] with $\gamma = 2$ and we have

$$2\Delta t\,S_1(h)\left\|\mathbf{v}_h^n\right\|_{0,\Omega}\left\|\nabla\mathbf{v}_h^n\right\|_{0,\Omega}\left\|\mathbf{v}_h^{n+1}-\mathbf{v}_h^n\right\|_{0,\Omega}$$
$$\leq 2(\Delta t)^2S_1^2(h)\left\|\mathbf{v}_h^n\right\|_{0,\Omega}^2\left\|\nabla\mathbf{v}_h^n\right\|_{0,\Omega}^2 + \frac{1}{2}\left\|\mathbf{v}_h^{n+1}-\mathbf{v}_h^n\right\|_{0,\Omega}^2.$$

We deduce

$$\left\|\mathbf{v}_h^{n+1}\right\|_{0,\Omega}^2 - \left\|\mathbf{v}_h^n\right\|_{0,\Omega}^2 + \left\|\mathbf{v}_h^{n+1} - \mathbf{v}_h^n\right\|_{0,\Omega}^2 + 2\Delta t\,\nu \left\|\nabla \mathbf{v}_h^{n+1}\right\|_{0,\Omega}^2$$
$$\leq \Delta t \frac{C_P^2}{\nu \rho^2} M^2 + \Delta t\,\nu \left\|\nabla \mathbf{v}_h^{n+1}\right\|_{0,\Omega}^2 + 2(\Delta t)^2 S_1^2(h) \left\|\mathbf{v}_h^n\right\|_{0,\Omega}^2 \left\|\nabla \mathbf{v}_h^n\right\|_{0,\Omega}^2$$
$$+\frac{1}{2}\left\|\mathbf{v}_h^{n+1} - \mathbf{v}_h^n\right\|_{0,\Omega}^2$$

which is equivalent to

$$\left\|\mathbf{v}_h^{n+1}\right\|_{0,\Omega}^2 - \left\|\mathbf{v}_h^n\right\|_{0,\Omega}^2 + \frac{1}{2}\left\|\mathbf{v}_h^{n+1} - \mathbf{v}_h^n\right\|_{0,\Omega}^2 + \Delta t\,\nu \left\|\nabla \mathbf{v}_h^{n+1}\right\|_{0,\Omega}^2$$
$$\leq \Delta t \frac{C_P^2}{\nu \rho^2} M^2 + 2(\Delta t)^2 S_1^2(h) \left\|\mathbf{v}_h^n\right\|_{0,\Omega}^2 \left\|\nabla \mathbf{v}_h^n\right\|_{0,\Omega}^2. \qquad [2.37]$$

We replace n with k in [2.37], and write the inequalities for $k = 0$

$$\left\|\mathbf{v}_h^1\right\|_{0,\Omega}^2 - \left\|\mathbf{v}_h^0\right\|_{0,\Omega}^2 + \frac{1}{2}\left\|\mathbf{v}_h^1 - \mathbf{v}_h^0\right\|_{0,\Omega}^2 + \Delta t\,\nu \left\|\nabla \mathbf{v}_h^1\right\|_{0,\Omega}^2$$
$$\leq \Delta t \frac{C_P^2}{\nu \rho^2} M^2 + 2(\Delta t)^2 S_1^2(h) \left\|\mathbf{v}_h^0\right\|_{0,\Omega}^2 \left\|\nabla \mathbf{v}_h^0\right\|_{0,\Omega}^2,$$

for $k = 1, \ldots, k = n$, and then we sum them. We have

$$\left\|\mathbf{v}_h^{n+1}\right\|_{0,\Omega}^2 - \left\|\mathbf{v}_h^0\right\|_{0,\Omega}^2 + \frac{1}{2}\sum_{k=0}^{n}\left\|\mathbf{v}_h^{k+1} - \mathbf{v}_h^k\right\|_{0,\Omega}^2 + \Delta t\,\nu \sum_{k=0}^{n}\left\|\nabla \mathbf{v}_h^{k+1}\right\|_{0,\Omega}^2$$
$$\leq (n+1)\Delta t \frac{C_P^2}{\nu \rho^2} M^2 + 2(\Delta t)^2 S_1^2(h) \sum_{k=0}^{n}\left\|\mathbf{v}_h^k\right\|_{0,\Omega}^2 \left\|\nabla \mathbf{v}_h^k\right\|_{0,\Omega}^2$$
$$\leq T \frac{C_P^2}{\nu \rho^2} M^2 + 2(\Delta t)^2 S_1^2(h) \sum_{k=0}^{n}\left\|\mathbf{v}_h^k\right\|_{0,\Omega}^2 \left\|\nabla \mathbf{v}_h^k\right\|_{0,\Omega}^2 \qquad [2.38]$$

which is equivalent to

$$\begin{aligned}
&\left\|\mathbf{v}_h^{n+1}\right\|_{0,\Omega}^2 + \frac{1}{2}\sum_{k=0}^{n}\left\|\mathbf{v}_h^{k+1} - \mathbf{v}_h^k\right\|_{0,\Omega}^2 + \Delta t\,\nu\left\|\nabla\mathbf{v}_h^{n+1}\right\|_{0,\Omega}^2\\
&+\Delta t\,\nu\sum_{k=1}^{n}\left\|\nabla\mathbf{v}_h^k\right\|_{0,\Omega}^2 \le T\frac{C_P^2}{\nu\rho^2}M^2 + \left\|\mathbf{v}_h^0\right\|_{0,\Omega}^2\\
&+2(\Delta t)^2 S_1^2(h)\left\|\mathbf{v}_h^0\right\|_{0,\Omega}^2\left\|\nabla\mathbf{v}_h^0\right\|_{0,\Omega}^2\\
&+2(\Delta t)^2 S_1^2(h)\sum_{k=1}^{n}\left\|\mathbf{v}_h^k\right\|_{0,\Omega}^2\left\|\nabla\mathbf{v}_h^k\right\|_{0,\Omega}^2. \qquad [2.39]
\end{aligned}$$

Given hypotheses [2.33] and [2.35], we have

$$\begin{aligned}
&T\frac{C_P^2}{\nu\rho^2}M^2 + \left\|\mathbf{v}_h^0\right\|_{0,\Omega}^2 + 2(\Delta t)^2 S_1^2(h)\left\|\mathbf{v}_h^0\right\|_{0,\Omega}^2\left\|\nabla\mathbf{v}_h^0\right\|_{0,\Omega}^2\\
&\le \frac{T\,C_P^2 M^2}{\nu\rho^2} + C_0^2 + 2\Delta t\frac{\nu}{4K}C_0^4
\end{aligned}$$

but, by construction $K \ge 1$, giving

$$\begin{aligned}
\frac{T\,C_P^2 M^2}{\nu\rho^2} + C_0^2 + 2\Delta t\frac{\nu}{4K}C_0^4 &\le \frac{T\,C_P^2 M^2}{\nu\rho^2} + C_0^2 + \Delta t\frac{\nu}{2}C_0^4\\
&\le \frac{T\,C_P^2 M^2}{\nu\rho^2} + C_0^2 + \frac{T\nu C_0^4}{2} \le K.
\end{aligned}$$

We deduce from [2.39] that

$$\begin{aligned}
&\left\|\mathbf{v}_h^{n+1}\right\|_{0,\Omega}^2 + \frac{1}{2}\sum_{k=0}^{n}\left\|\mathbf{v}_h^{k+1} - \mathbf{v}_h^k\right\|_{0,\Omega}^2 + \Delta t\,\nu\left\|\nabla\mathbf{v}_h^{n+1}\right\|_{0,\Omega}^2\\
&+\Delta t\sum_{k=1}^{n}\left(\nu - 2\Delta t\,S_1^2(h)\left\|\mathbf{v}_h^k\right\|_{0,\Omega}^2\right)\left\|\nabla\mathbf{v}_h^k\right\|_{0,\Omega}^2 \le K. \qquad [2.40]
\end{aligned}$$

We shall prove by induction that $\left\|\mathbf{v}_h^k\right\|_{0,\Omega}^2 \leq K$, $1 \leq k$. For $n = 0$ in [2.37] and preceding expressions, we have

$$\left\|\mathbf{v}_h^1\right\|_{0,\Omega}^2 + \frac{1}{2}\left\|\mathbf{v}_h^1 - \mathbf{v}_h^0\right\|_{0,\Omega}^2 + \Delta t\, \nu \left\|\nabla \mathbf{v}_h^1\right\|_{0,\Omega}^2$$

$$\leq \Delta t \frac{C_P^2}{\nu \rho^2} M^2 + \left\|\mathbf{v}_h^0\right\|_{0,\Omega}^2 + 2(\Delta t)^2 S_1^2(h) \left\|\mathbf{v}_h^0\right\|_{0,\Omega}^2 \left\|\nabla \mathbf{v}_h^0\right\|_{0,\Omega}^2 \leq K$$

which implies $\left\|\mathbf{v}_h^1\right\|_{0,\Omega}^2 \leq K$.

We assume that $\left\|\mathbf{v}_h^k\right\|_{0,\Omega}^2 \leq K$, $1 \leq k \leq n$. Under hypothesis [2.35], we have

$$2\Delta t\, S_1^2(h) \left\|\mathbf{v}_h^k\right\|_{0,\Omega}^2 \leq 2\frac{\nu}{4K} K \leq \frac{\nu}{2}, \quad 1 \leq k \leq n.$$

Thus,

$$\nu - 2\Delta t\, S_1^2(h) \left\|\mathbf{v}_h^k\right\|_{0,\Omega}^2 \geq \frac{\nu}{2}, \quad 1 \leq k \leq n.$$

We obtain from [2.40] that

$$\left\|\mathbf{v}_h^{n+1}\right\|_{0,\Omega}^2 + \frac{1}{2}\sum_{k=0}^{n} \left\|\mathbf{v}_h^{k+1} - \mathbf{v}_h^k\right\|_{0,\Omega}^2 + \Delta t\, \nu \left\|\nabla \mathbf{v}_h^{n+1}\right\|_{0,\Omega}^2$$

$$+\Delta t \frac{\nu}{2} \sum_{k=1}^{n} \left\|\nabla \mathbf{v}_h^k\right\|_{0,\Omega}^2 \leq K \qquad [2.41]$$

giving $\left\|\mathbf{v}_h^{n+1}\right\|_{0,\Omega}^2 \leq K$, thus $\left\|\mathbf{v}_h^k\right\|_{0,\Omega}^2 \leq K$, $1 \leq k$. We have seen that $\left\|\mathbf{v}_h^k\right\|_{0,\Omega}^2 \leq K$, $1 \leq k \leq n$ implies [2.41], which in turn gives [2.36]. □

We shall now discuss inequality [2.30], which is the inverse of the Poincaré inequality in a space of finite dimensions.

We assume that Ω is a polygonal set and let $(\mathcal{T}_h)_{h>0}$ be a family of triangulations for Ω.

To approximate the velocity, we can use, for example, the finite element $\mathbb{P}_1 +$ *bubble*. We obtain a vector space of finite dimensions, which is as follows:

$$W_h = \left\{\mathbf{w}_h \in \left(C^0(\overline{\Omega})\right)^2;\ \mathbf{w}_h = 0 \text{ on } \partial\Omega,\ \forall T \in \mathcal{T}_h,\ \mathbf{w}_{h|T} \in \mathbb{P}_1 + bubble\right\}.$$

Since $\mathbf{w}_h \in W_h$ is continuous in $\overline{\Omega}$, due to Theorem 1.4-3, p. 27, [RAV 98], we have $\mathbf{w}_h \in (H^1(\Omega))^2$ and, taking account of the boundary conditions, we obtain $W_h \subset W$.

In the previous chapter, we saw Definitions 1.2 and 1.3 of a regular and uniformly regular triangulation respectively.

In light of Corollary 1.141, p.76, [ERN 04], if the triangulation is uniformly regular (also known as quasi-uniform), we have the following inequality:

$$||\mathbf{w}_h||_{1,\Omega} \leq C\,h^{-1}\,||\mathbf{w}_h||_{0,\Omega}$$

giving [2.30] with $S(h) = C\,h^{-1}$.

We shall now give a useful characterisation of $S(h)$ as a function of its eigenvalues. Let $m = dim(W_h)$ and $\{\varphi_h^i,\ i = 1, \ldots, m\}$ be a basis of W_h. There exist the eigenvalues

$$0 < \lambda_h^1 \leq \lambda_h^2 \leq \cdots \leq \lambda_h^m$$

and the eigenvectors $\{\psi_h^i \in W_h,\ i = 1, \ldots, m\}$ such that

$$\left(\nabla\psi_h^i, \nabla\mathbf{w}_h\right)_{0,\Omega} = \lambda_h^i \left(\psi_h^i, \mathbf{w}_h\right)_{0,\Omega}, \quad \mathbf{w}_h \in W_h \qquad [2.42]$$

and

$$\left(\psi_h^i, \psi_h^j\right)_{0,\Omega} = \delta_{ij}. \qquad [2.43]$$

The eigenvectors form an orthonormal basis of W_h for the scalar product $(\cdot,\cdot)_{0,\Omega}$.

We can construct the matrices

$$K = \left(\left(\nabla\varphi_h^j, \nabla\varphi_h^i\right)_{0,\Omega}\right)_{0\leq i,j\leq m}, \quad M = \left(\left(\varphi_h^j, \varphi_h^i\right)_{0,\Omega}\right)_{0\leq i,j\leq m}$$

and define the Rayleigh quotient $\mathcal{R} : \mathbb{R}^m \setminus \{0\} \to \mathbb{R}$ by

$$\mathcal{R}(\xi) = \frac{\langle K\xi, \xi\rangle_{\mathbb{R}^m}}{\langle M\xi, \xi\rangle_{\mathbb{R}^m}}$$

where $\langle \cdot, \cdot \rangle_{\mathbb{R}^m}$ is the Euclidean scalar product in $\mathbb{R}^m$. We obtain the following result:

$$\mathcal{R}(\mathbb{R}^m \setminus \{0\}) = [\lambda_h^1, \lambda_h^m]$$

i.e.

$$\lambda_h^1 \leq \frac{\langle K\boldsymbol{\xi}, \boldsymbol{\xi} \rangle_{\mathbb{R}^m}}{\langle M\boldsymbol{\xi}, \boldsymbol{\xi} \rangle_{\mathbb{R}^m}} \leq \lambda_h^m, \quad \forall \boldsymbol{\xi} \in \mathbb{R}^m \setminus \{0\}$$

and the limits λ_h^1 and λ_h^m are optimal. For $\mathbf{w}_h = \sum_{i=1}^m \xi_i \boldsymbol{\varphi}_h^i$ and $\boldsymbol{\xi} = (\xi_1, \ldots, \xi_m)^T$, we have

$$\frac{(\nabla \mathbf{w}_h, \nabla \mathbf{w}_h)_{0,\Omega}}{(\mathbf{w}_h, \mathbf{w}_h)_{0,\Omega}} = \frac{\langle K\boldsymbol{\xi}, \boldsymbol{\xi} \rangle_{\mathbb{R}^m}}{\langle M\boldsymbol{\xi}, \boldsymbol{\xi} \rangle_{\mathbb{R}^m}} \leq \lambda_h^m.$$

Thus,

$$||\nabla \mathbf{w}_h||_{0,\Omega}^2 \leq \lambda_h^m \, ||\mathbf{w}_h||_{0,\Omega}^2$$

from which we can take $S(h) = \sqrt{\lambda_h^m}$.

2.5. Some properties of the trilinear forms c and $\tilde{c}$

All the results in this section are obtained under the assumption that $\Omega \subset \mathbb{R}^2$ is an open, connected, and bounded set with Lipschitz boundary. We recall the notations $W = (H_0^1(\Omega))^2$, $Q = L_0^2(\Omega)$, $V = \{\mathbf{w} \in (H_0^1(\Omega))^2;\ \nabla \cdot \mathbf{w} = 0\}$,

$$c : W \times W \times W \to \mathbb{R}, \quad c(\mathbf{u}, \mathbf{v}, \mathbf{w}) = \int_\Omega [(\mathbf{u} \cdot \nabla)\mathbf{v}] \cdot \mathbf{w}\, d\mathbf{x}$$

$$\tilde{c} : W \times W \times W \to \mathbb{R}, \quad \tilde{c}(\mathbf{u}, \mathbf{v}, \mathbf{w}) = \frac{1}{2} c(\mathbf{u}, \mathbf{v}, \mathbf{w}) - \frac{1}{2} c(\mathbf{u}, \mathbf{w}, \mathbf{v}).$$

LEMMA 2.1.– *The following equality holds*

$$\forall \mathbf{u} \in V, \forall \mathbf{w} \in W, \quad c(\mathbf{u}, \mathbf{w}, \mathbf{w}) = 0. \qquad [2.44]$$

DEMONSTRATION 2.3.– We have

$$
\begin{aligned}
c(\mathbf{u}, \mathbf{w}, \mathbf{w}) &= \int_\Omega [(\mathbf{u} \cdot \nabla)\mathbf{w}] \cdot \mathbf{w}\, d\mathbf{x} \\
&= \int_\Omega \left(u_1 \frac{\partial w_1}{\partial x_1} w_1 + u_2 \frac{\partial w_1}{\partial x_2} w_1 + u_1 \frac{\partial w_2}{\partial x_1} w_2 + u_2 \frac{\partial w_2}{\partial x_2} w_2 \right) d\mathbf{x} \\
&= \int_\Omega \frac{1}{2} \left(u_1 \frac{\partial (w_1)^2}{\partial x_1} + u_2 \frac{\partial (w_1)^2}{\partial x_2} + u_1 \frac{\partial (w_2)^2}{\partial x_1} + u_2 \frac{\partial (w_2)^2}{\partial x_2} \right) d\mathbf{x} \\
&= - \int_\Omega \frac{1}{2} \left(\frac{\partial u_1}{\partial x_1} (w_1)^2 + \frac{\partial u_2}{\partial x_2} (w_1)^2 + \frac{\partial u_1}{\partial x_1} (w_2)^2 + \frac{\partial u_2}{\partial x_2} (w_2)^2 \right) d\mathbf{x} \\
&= -\frac{1}{2} \int_\Omega (\nabla \cdot \mathbf{u}) \left((w_1)^2 + (w_2)^2 \right) d\mathbf{x} = 0.
\end{aligned}
$$

□

LEMMA 2.2.– *If* **u**, **v**, **w** *are in* W, *then*

$$
|c(\mathbf{u}, \mathbf{v}, \mathbf{w})| \leq \sqrt{2}\, ||\mathbf{u}||_{0,\Omega}^{1/2} \, ||\nabla \mathbf{u}||_{0,\Omega}^{1/2} \, ||\nabla \mathbf{v}||_{0,\Omega} \, ||\mathbf{w}||_{0,\Omega}^{1/2} \, ||\nabla \mathbf{w}||_{0,\Omega}^{1/2} . \qquad [2.45]
$$

DEMONSTRATION 2.4.– We have

$$
c(\mathbf{u}, \mathbf{v}, \mathbf{w}) = \sum_{i,j=1}^{2} \int_\Omega u_i \frac{\partial v_j}{\partial x_i} w_j \, d\mathbf{x}.
$$

For $\Omega \subset \mathbb{R}^2$, we have the inclusion $H^1(\Omega) \subset L^4(\Omega)$ and using Hölder's inequality [A.3], we have

$$
\left| \int_\Omega u_i \frac{\partial v_j}{\partial x_i} w_j \, d\mathbf{x} \right| \leq ||u_i||_{L^4(\Omega)} \left\| \frac{\partial v_j}{\partial x_i} \right\|_{L^2(\Omega)} \left\| w_j \right\|_{L^4(\Omega)}
$$

giving

$$
\begin{aligned}
|c(\mathbf{u}, \mathbf{v}, \mathbf{w})| &\leq \sum_{i,j=1}^{2} ||u_i||_{L^4(\Omega)} \left\| \frac{\partial v_j}{\partial x_i} \right\|_{L^2(\Omega)} \left\| w_j \right\|_{L^4(\Omega)} \\
&= \sum_{i,j=1}^{2} \left(||u_i||_{L^4(\Omega)} \left\| w_j \right\|_{L^4(\Omega)} \right) \left\| \frac{\partial v_j}{\partial x_i} \right\|_{L^2(\Omega)} .
\end{aligned}
$$

We shall use the Cauchy-Schwarz inequality in $\mathbb{R}^n$ (see [A.2])

$$\left|\sum_{k=1}^{n} \alpha_k \beta_k\right| \leq \left(\sum_{k=1}^{n} \alpha_k^2\right)^{1/2} \left(\sum_{k=1}^{n} \beta_k^2\right)^{1/2}$$

with $\alpha_k = \|u_i\|_{L^4(\Omega)} \left\|w_j\right\|_{L^4(\Omega)}$ and $\beta_k = \left\|\frac{\partial v_j}{\partial x_i}\right\|_{L^2(\Omega)}$, giving

$$\begin{aligned}
&\sum_{i,j=1}^{2} \left(\|u_i\|_{L^4(\Omega)} \left\|w_j\right\|_{L^4(\Omega)}\right) \left\|\frac{\partial v_j}{\partial x_i}\right\|_{L^2(\Omega)} \\
&\leq \left(\sum_{i,j=1}^{2} \|u_i\|_{L^4(\Omega)}^2 \left\|w_j\right\|_{L^4(\Omega)}^2\right)^{1/2} \left(\sum_{i,j=1}^{2} \left\|\frac{\partial v_j}{\partial x_i}\right\|_{L^2(\Omega)}\right)^{1/2} \\
&= \left(\left(\sum_{i=1}^{2} \|u_i\|_{L^4(\Omega)}^2\right)\left(\sum_{j=1}^{2} \left\|w_j\right\|_{L^4(\Omega)}^2\right)\right)^{1/2} \|\nabla \mathbf{v}\|_{0,\Omega}.
\end{aligned}$$

We shall use Lemma 3.3, p. 197 [TEM 01b], which gives the expression

$$\|u_i\|_{L^4(\Omega)} \leq 2^{1/4} \|u_i\|_{L^2(\Omega)}^{1/2} \|\nabla u_i\|_{L^2(\Omega)}^{1/2}.$$

Thus,

$$\begin{aligned}
&\|u_1\|_{L^4(\Omega)}^2 + \|u_2\|_{L^4(\Omega)}^2 \\
&\leq \sqrt{2}\,\|u_1\|_{L^2(\Omega)} \|\nabla u_1\|_{L^2(\Omega)} + \sqrt{2}\,\|u_2\|_{L^2(\Omega)} \|\nabla u_2\|_{L^2(\Omega)} \\
&\leq \sqrt{2}\left(\|u_1\|_{L^2(\Omega)}^2 + \|u_2\|_{L^2(\Omega)}^2\right)^{1/2} \left(\|\nabla u_1\|_{L^2(\Omega)}^2 + \|\nabla u_2\|_{L^2(\Omega)}^2\right)^{1/2} \\
&= \sqrt{2}\,\|\mathbf{u}\|_{0,\Omega} \|\nabla \mathbf{u}\|_{0,\Omega}.
\end{aligned}$$

In going from the second to the third line above, we have used the Cauchy-Schwarz inequality in $\mathbb{R}^n$ (see [A.2]). Similarly, we have

$$\|w_1\|_{L^4(\Omega)}^2 + \|w_2\|_{L^4(\Omega)}^2 \leq \sqrt{2}\,\|\mathbf{w}\|_{0,\Omega} \|\nabla \mathbf{w}\|_{0,\Omega}.$$

Finally, we obtain the following equation:

$$
\begin{aligned}
|c(\mathbf{u},\mathbf{v},\mathbf{w})| &\leq \left(\sqrt{2}\,\|\mathbf{u}\|_{0,\Omega}\,\|\nabla\mathbf{u}\|_{0,\Omega}\right)^{1/2}\left(\sqrt{2}\,\|\mathbf{w}\|_{0,\Omega}\,\|\nabla\mathbf{w}\|_{0,\Omega}\right)^{1/2}\|\nabla\mathbf{v}\|_{0,\Omega} \\
&= \sqrt{2}\,\|\mathbf{u}\|_{0,\Omega}^{1/2}\,\|\nabla\mathbf{u}\|_{0,\Omega}^{1/2}\,\|\nabla\mathbf{v}\|_{0,\Omega}\,\|\mathbf{w}\|_{0,\Omega}^{1/2}\,\|\nabla\mathbf{w}\|_{0,\Omega}^{1/2}.
\end{aligned}
$$

□

LEMMA 2.3.– *There exists $M_c > 0$ such that*

$$\forall \mathbf{u},\mathbf{v},\mathbf{w} \in W, \quad |c(\mathbf{u},\mathbf{v},\mathbf{w})| \leq M_c\,\|\mathbf{u}\|_{1,\Omega}\,\|\mathbf{v}\|_{1,\Omega}\,\|\mathbf{w}\|_{1,\Omega}. \qquad [2.46]$$

DEMONSTRATION 2.5.–

Solution 1. As for the demonstration of lemma 2.2, we have

$$|c(\mathbf{u},\mathbf{v},\mathbf{w})| \leq \left(\sum_{i=1}^{2}\|u_i\|_{L^4(\Omega)}^2\right)^{1/2}\left(\sum_{j=1}^{2}\|w_j\|_{L^4(\Omega)}^2\right)^{1/2}\|\nabla\mathbf{v}\|_{0,\Omega}.$$

For $\Omega \subset \mathbb{R}^2$ bounded with Lipschitz boundary, the injection $H^1(\Omega) \subset L^4(\Omega)$ is continuous (see for example [BRÉ 05], p.168). Thus

$$\forall u \in H^1(\Omega),\ \|u\|_{L^4(\Omega)} \leq C\,\|u\|_{H^1(\Omega)}$$

and we obtain the following equation:

$$|c(\mathbf{u},\mathbf{v},\mathbf{w})| \leq C^2\,\|\mathbf{u}\|_{1,\Omega}\,\|\mathbf{v}\|_{1,\Omega}\,\|\mathbf{w}\|_{1,\Omega}.$$

Solution 2. We shall use lemma 2.2 and the Poincaré inequality

$$\exists C_P > 0, \forall u_i \in H_0^1(\Omega),\ \|u_i\|_{0,\Omega} \leq C_P\,\|\nabla u_i\|_{0,\Omega}.$$

We have an identical inequality for the vector functions

$$\forall \mathbf{u} \in W,\ \|\mathbf{u}\|_{0,\Omega} \leq C_P\,\|\nabla\mathbf{u}\|_{0,\Omega}.$$

Using lemma 2.45, we have

$$
\begin{aligned}
|c(\mathbf{u},\mathbf{v},\mathbf{w})| &\leq \sqrt{2}C_P\,\|\nabla\mathbf{u}\|_{0,\Omega}\,\|\nabla\mathbf{v}\|_{0,\Omega}\,\|\nabla\mathbf{w}\|_{0,\Omega} \\
&\leq \sqrt{2}C_P\,\|\mathbf{u}\|_{1,\Omega}\,\|\mathbf{v}\|_{1,\Omega}\,\|\mathbf{w}\|_{1,\Omega}.
\end{aligned}
$$

□

LEMMA 2.4.– *The following equality holds*

$$\forall \mathbf{u} \in V, \forall \mathbf{v}, \mathbf{w} \in W, \quad \tilde{c}(\mathbf{u}, \mathbf{v}, \mathbf{w}) = c(\mathbf{u}, \mathbf{v}, \mathbf{w}). \qquad [2.47]$$

DEMONSTRATION 2.6.– If $\mathbf{u} \in V$ and $\mathbf{v}, \mathbf{w} \in W$, due to [2.44] we have

$$\begin{aligned} 0 &= c(\mathbf{u}, \mathbf{v} + \mathbf{w}, \mathbf{v} + \mathbf{w}) \\ &= c(\mathbf{u}, \mathbf{v}, \mathbf{v}) + c(\mathbf{u}, \mathbf{w}, \mathbf{w}) + c(\mathbf{u}, \mathbf{v}, \mathbf{w}) + c(\mathbf{u}, \mathbf{w}, \mathbf{v}) \\ &= c(\mathbf{u}, \mathbf{v}, \mathbf{w}) + c(\mathbf{u}, \mathbf{w}, \mathbf{v}). \end{aligned}$$

By replacing $-c(\mathbf{u}, \mathbf{w}, \mathbf{v})$ with $c(\mathbf{u}, \mathbf{v}, \mathbf{w})$ in the definition of $\tilde{c}$, we obtain the specified conclusion. □

LEMMA 2.5.– *The following inequality holds*

$$\forall \mathbf{u}, \mathbf{v}, \mathbf{w} \in W, \quad |\tilde{c}(\mathbf{u}, \mathbf{v}, \mathbf{w})| \leq M_c \, ||\mathbf{u}||_{1,\Omega} \, ||\mathbf{v}||_{1,\Omega} \, ||\mathbf{w}||_{1,\Omega} \, . \qquad [2.48]$$

DEMONSTRATION 2.7.–

$$\begin{aligned} |\tilde{c}(\mathbf{u}, \mathbf{v}, \mathbf{w})| &= \left| \frac{1}{2} c(\mathbf{u}, \mathbf{v}, \mathbf{w}) - \frac{1}{2} c(\mathbf{u}, \mathbf{w}, \mathbf{v}) \right| \\ &\leq \frac{1}{2} \left(|c(\mathbf{u}, \mathbf{v}, \mathbf{w})| + |c(\mathbf{u}, \mathbf{w}, \mathbf{v})| \right) \\ &\leq M_c \, ||\mathbf{u}||_{1,\Omega} \, ||\mathbf{v}||_{1,\Omega} \, ||\mathbf{w}||_{1,\Omega} \, . \end{aligned}$$

□

2.6. Numerical tests: flow around a disc

We have adapted the 2D benchmark found in [SCH 96]. We study Navier-Stokes flow in a channel of length $L = 2.2\,m$ and height $H = 0.41\,m$ around a disc centered at $(0.2, 0.2)$ and radius $r = 0.05\,m$ (see Figure 2.1).

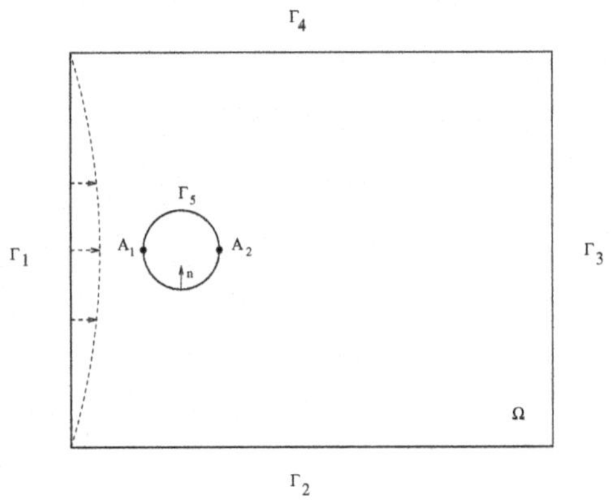

Figure 2.1. *Geometric configuration*

The fluid has mass density $\rho = 1\, kg/(m^3)$ and dynamic viscosity $\mu = 0.001\, Pa\, s$. There are no external body forces, $\mathbf{f} = (0, 0)$. We work with the boundary conditions:

$$\begin{aligned} \mathbf{v} &= \mathbf{g}, \text{ on } \Gamma_1 \times]0, T[\\ \mathbf{v} &= 0, \text{ on } (\Gamma_2 \cup \Gamma_4 \cup \Gamma_5) \times]0, T[\\ \sigma \mathbf{n} &= 0, \text{ on } \Gamma_3 \times]0, T[\end{aligned}$$

where

$$\mathbf{g}(x_1, x_2, t) = \begin{cases} \left(1.5\, \overline{U} \frac{x_2(H-x_2)}{(H/2)^2} \frac{(1-\cos(\pi t))}{2},\ 0\right), & (x_1, x_2) \in \Sigma_1, 0 \leq t \leq 1 \\ \left(1.5\, \overline{U} \frac{x_2(H-x_2)}{(H/2)^2},\ 0\right), & (x_1, x_2) \in \Sigma_1, 1 < t \leq T \end{cases}$$

$\overline{U} = 1\, m/s$, $T = 10\, s$ and $\mathbf{n}$ is the unit vector, normal to the boundary, pointing out of the fluid domain. We work with the initial condition $\mathbf{v}_0 = 0$.

We want to calculate the time evolution:

– of the difference in pressure between the points $A_1 = (0.2 - r, 0.2)$ and $A_2 = (0.2 + r, 0.2)$, $\Delta P(t) = p(A_1, t) - p(A_2, t)$,

– of the drag coefficient $c_D(t)$ and the lift coefficient $c_L(t)$, given in the case of the disc by

$$c_D(t) = \frac{2F_D(t)}{\overline{U}^2 2r\rho}, \quad c_L(t) = \frac{2F_L(t)}{\overline{U}^2 2r\rho}, \quad (F_D(t), F_L(t))^T = -\int_{\Gamma_5} \sigma \mathbf{n}\, ds.$$

2.6.1. *Semi-implict treatment of the nonlinear term*

We have used the time step $\Delta t = 0.01$ s and three meshes with the properties given in Table 2.1. We use the finite elements $\mathbb{P}_1$ + *bubble* for the velocity and $\mathbb{P}_1$ for the pressure.

	Vertices	Triangles
mesh1	1094	2008
mesh2	1981	3722
mesh3	3004	5708

Table 2.1. *The number of vertices and triangles*

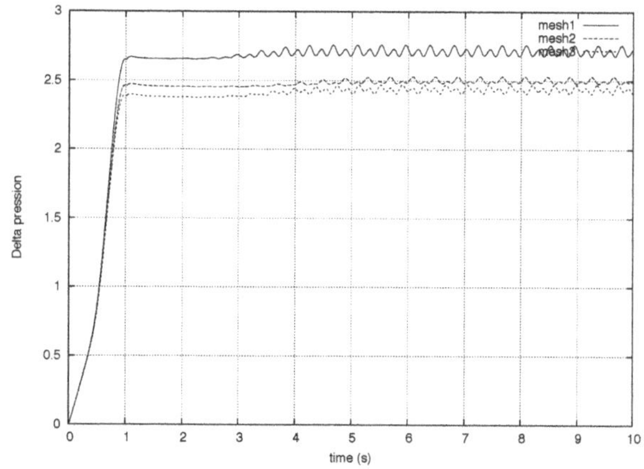

Figure 2.2. *The pressure difference between points A_1 and A_2 for the semi-implicit scheme with $\Delta t = 0.01$ and three meshes*

The flow becomes periodic after $t = 4\,s$, see Figures 2.2 and 2.3. The velocity and the pressure are presented in Figure 2.4. We have also conducted the tests with three values for Δt : 0.001, 0.05, and 0.01 (see Figures 2.5 and 2.6).

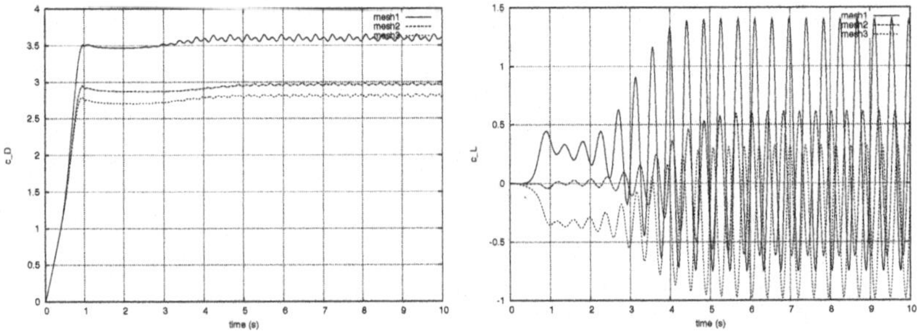

Figure 2.3. *The drag (left) and lift (right) coefficients for the semi-implicit scheme with* $\Delta t = 0.01$ *and three meshes*

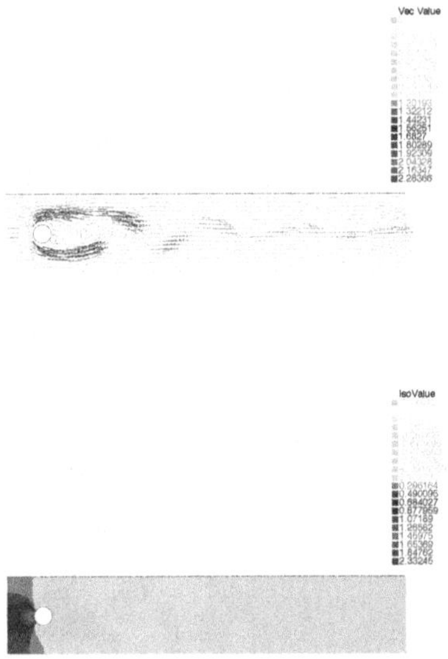

Figure 2.4. *The velocity (above) and the pressure (below) at time* $t = 10\ s$ *with mesh2 and* $\Delta t = 0.01$ *for the semi-implicit scheme. For a color version of this figure, see www.iste.co.uk/murea/schemes.zip*

2.6.2. *Implicit treatment of the nonlinear term*

We have used the time step $\Delta t = 0.01\ s$ and the mesh *mesh2* (see Table 2.1). We use the finite elements $\mathbb{P}_1$ + *bubble* for velocity and $\mathbb{P}_1$ for pressure. For Newton's method, we have set $tol = 10^{-5}$ for the stopping criterion.

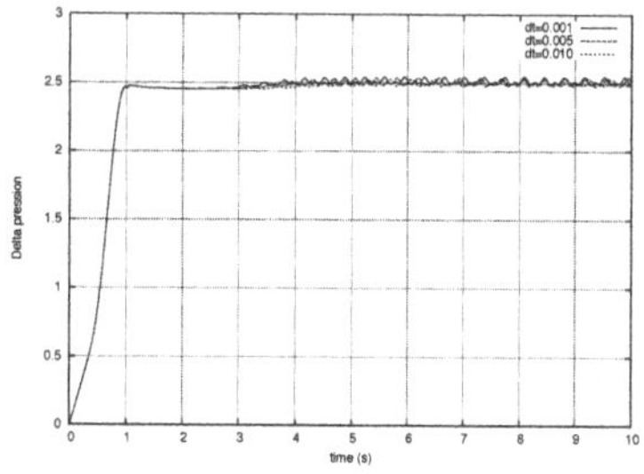

Figure 2.5. *The pressure difference between points A_1 and A_2 for the semi-implicit scheme with mesh2 and three different values for Δt*

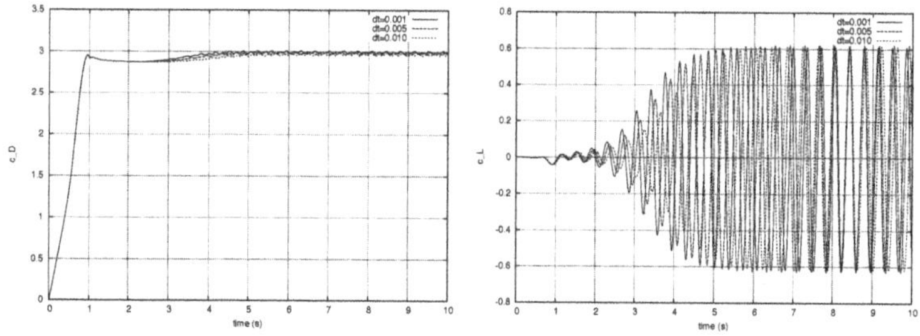

Figure 2.6. *The drag (left) and lift (right) coefficients for the semi-implicit scheme with mesh2 and three different values for Δt*

To initialize Newton's method at time $t = (n + 1)\Delta t$ we use:

– for $0 \leq t \leq 1$, the solution of a Stokes problem with boundary conditions:

$$\mathbf{v} = \mathbf{g}, \text{ on } \Gamma_1 \times]0, T[$$

$$\mathbf{v} = 0, \text{ on } (\Gamma_2 \cup \Gamma_4 \cup \Gamma_5) \times]0, T[$$

$$\sigma \mathbf{n} = 0, \text{ on } \Gamma_3 \times]0, T[$$

– for $1 < t$, the solution at time $t = n\Delta t$.

On average, Newton's method carries out 3 iterations for each step in time. The relationship between the computation times is

$$\frac{CPU implicit}{CPU semi-implicit} = 3.56.$$

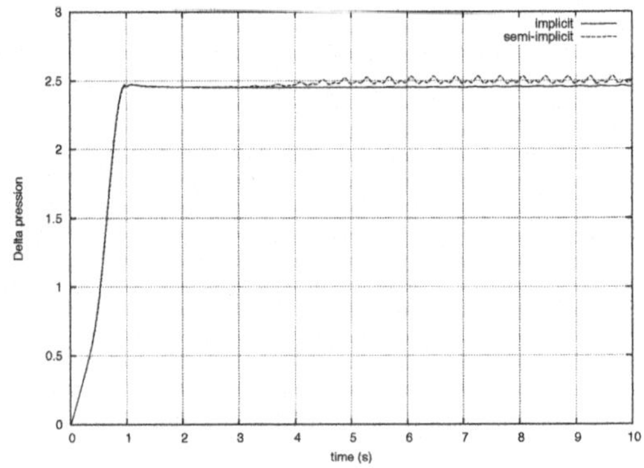

Figure 2.7. *The pressure difference between points A_1 and A_2 for the implicit and semi-implicit schemes with $\Delta t = 0.01$*

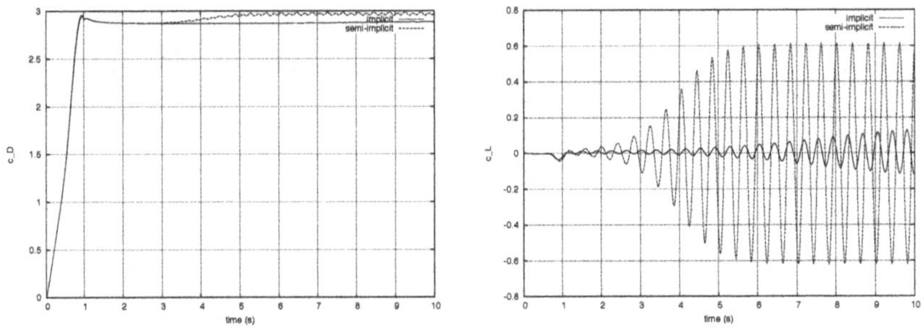

Figure 2.8. *The drag (left) and lift (right) coefficients for the implicit and semi-implicit schemes with $\Delta t = 0.01$*

The numerical results obtained by the implicit and semi-implicit schemes are compared in Figures 2.7 and 2.8. The velocity and pressure obtained by the implicit scheme are presented in Figure 2.9.

2.6.3. *Explicit treatment of the nonlinear term*

In order to simplify the numerical implementation, we shall use c in the place of $\tilde{c}$ in scheme [2.27]–[2.28]. We present only the scheme that is semi-discretized in time.

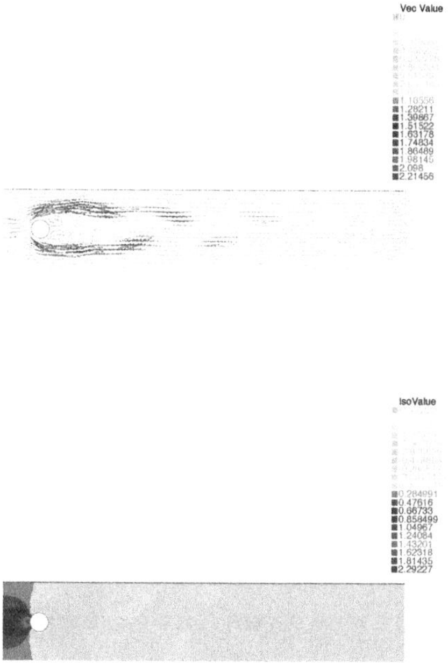

Figure 2.9. *The velocity (above) and pressure (below) at time $t = 10\,s$ with mesh2 for the implicit scheme. For a color version of this figure, see www.iste.co.uk/murea/schemes.zip*

Find $\mathbf{v}^{n+1} \in (H^1(\Omega))^2$, $\mathbf{v}^{n+1} = \mathbf{g}$ on Γ_1, $\mathbf{v}^{n+1} = 0$ on $\Gamma_2 \cup \Gamma_4 \cup \Gamma_5$ and $p^{n+1} \in L^2(\Omega)$, such that

$$\rho\frac{1}{\Delta t}\int_\Omega \mathbf{v}^{n+1}\cdot\mathbf{w}\,d\mathbf{x} + a(\mathbf{v}^{n+1},\mathbf{w}) + b(\mathbf{w},p^{n+1})$$

$$= \int_\Omega \mathbf{f}^{n+1}\cdot\mathbf{w}\,d\mathbf{x} + \rho\frac{1}{\Delta t}\int_\Omega \mathbf{v}^{n}\cdot\mathbf{w}\,d\mathbf{x} - \rho c(\mathbf{v}^n,\mathbf{v}^n,\mathbf{w}), \quad \forall \mathbf{w}\in W$$

$$b(\mathbf{v}^{n+1},q) = 0, \quad \forall q \in Q$$

where $W = \left\{\mathbf{w}\in (H^1(\Omega))^2;\ \mathbf{w} = 0 \text{ on } \Gamma_1\cup\Gamma_2\cup\Gamma_4\cup\Gamma_5\right\}$ and $Q = L^2(\Omega)$. We initialize with $\mathbf{v}^0 = \mathbf{v}_0$.

This scheme is of order one in time. To prove this, we assume that

$$\mathbf{v} \in C^2\left(]0,T[,(H^1(\Omega))^2\right)$$

and we set

$$r = \rho \int_\Omega \frac{\mathbf{v}(t_{n+1}) - \mathbf{v}(t_n)}{\Delta t} \cdot \mathbf{w}\, d\mathbf{x} + \rho c(\mathbf{v}(t_n), \mathbf{v}(t_n), \mathbf{w})$$

$$+a(\mathbf{v}(t_{n+1}), \mathbf{w}) + b(\mathbf{w}, p(t_{n+1})) - \int_\Omega \mathbf{f}(t_{n+1}) \cdot \mathbf{w}\, d\mathbf{x}.$$

Taking account of [2.5] for $t = t_{n+1}$, we obtain the following equation:

$$r = \rho \int_\Omega \frac{\mathbf{v}(t_{n+1}) - \mathbf{v}(t_n)}{\Delta t} \cdot \mathbf{w}\, d\mathbf{x} + \rho c(\mathbf{v}(t_n), \mathbf{v}(t_n), \mathbf{w})$$

$$-\rho \int_\Omega \mathbf{v}'(t_{n+1}) \cdot \mathbf{w}\, d\mathbf{x} - \rho c(\mathbf{v}(t_{n+1}), \mathbf{v}(t_{n+1}), \mathbf{w})$$

$$= \rho \int_\Omega \left(\frac{\mathbf{v}(t_{n+1}) - \mathbf{v}(t_n)}{\Delta t} - \mathbf{v}'(t_{n+1}) \right) \cdot \mathbf{w}\, d\mathbf{x}$$

$$+\rho c(\mathbf{v}(t_n), \mathbf{v}(t_n), \mathbf{w}) - \rho c(\mathbf{v}(t_{n+1}), \mathbf{v}(t_{n+1}), \mathbf{w}).$$

However, using Taylor's formula, we have

$$\mathbf{v}(t_{n+1}) = \mathbf{v}(t_n) + (\Delta t)\mathbf{v}'(t_{n+1}) + \frac{(\Delta t)^2}{2}\mathbf{v}''(\tau), \quad \tau \in]t_n, t_{n+1}[.$$

Thus

$$\rho \int_\Omega \left(\frac{\mathbf{v}(t_{n+1}) - \mathbf{v}(t_n)}{\Delta t} - \mathbf{v}'(t_{n+1}) \right) \cdot \mathbf{w}\, d\mathbf{x} = \Delta t\, \rho \int_\Omega \frac{1}{2}\mathbf{v}''(\tau) \cdot \mathbf{w}\, d\mathbf{x}$$

and

$$c\left(\mathbf{v}(t_{n+1}), \mathbf{v}(t_{n+1}), \mathbf{w}\right) = c\left(\mathbf{v}(t_n), \mathbf{v}(t_n), \mathbf{w}\right)$$

$$+(\Delta t)c\left(\mathbf{v}(t_n), \mathbf{v}'(t_{n+1}) + \frac{\Delta t}{2}\mathbf{v}''(\tau), \mathbf{w}\right)$$

$$+(\Delta t)c\left(\mathbf{v}'(t_{n+1}) + \frac{\Delta t}{2}\mathbf{v}''(\tau), \mathbf{v}(t_n), \mathbf{w}\right)$$

$$+(\Delta t)^2 c\left(\mathbf{v}'(t_{n+1}) + \frac{\Delta t}{2}\mathbf{v}''(\tau), \mathbf{v}'(t_{n+1}) + \frac{\Delta t}{2}\mathbf{v}''(\tau), \mathbf{w}\right)$$

giving $r = O(\Delta t)$.

We have used the mesh *mesh2* (see Table 2.1) with the finite elements $\mathbb{P}_1$ + *bubble* for velocity and $\mathbb{P}_1$ for pressure. In our example, the explicit scheme does not work with $\Delta t = 0.01$ or $\Delta t = 0.005$. We have done the tests with $\Delta t = 0.001$.

At each step in time, we solve a symmetric linear system in the explicit scheme, but a non-symmetric system in the semi-implicit scheme. We can use higher-performance solvers for the symmetric linear systems. However, the explicit scheme is stable provided that the time step is small.

The numerical results obtained by the explicit scheme are compared with the semi-implicit scheme in Figures 2.10 and 2.11.

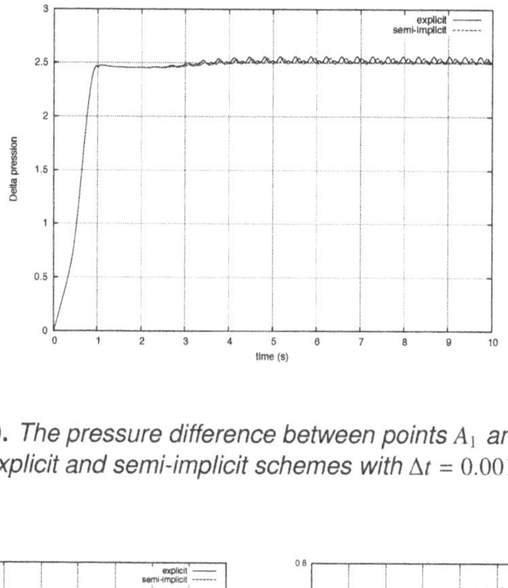

Figure 2.10. *The pressure difference between points A_1 and A_2 for the explicit and semi-implicit schemes with $\Delta t = 0.001$*

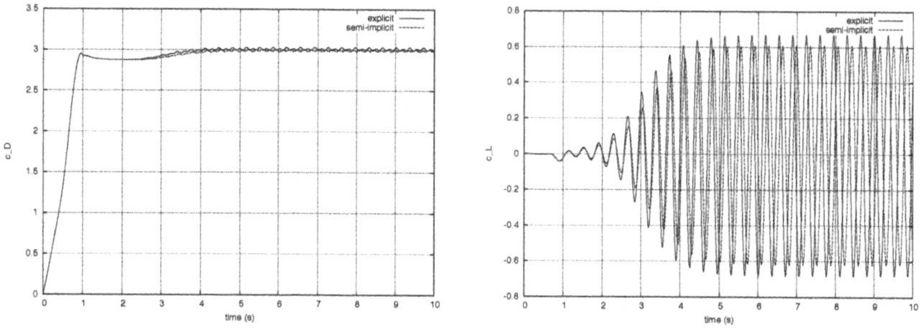

Figure 2.11. *The drag (left) and lift (right) coefficients for the explicit and semi-implicit schemes with $\Delta t = 0.001$*

We can calculate $S(h)$ numerically from inequality [2.30] using eigenvalues. For the meshes in Table 2.1 we have

	h	S(h)
mesh1	0.051928	544.013
mesh2	0.0372083	723.499
mesh3	0.0304521	944.319

Table 2.2. *The value $S(h)$ as a function of the mesh size h*

We can see the graph of $S(h)$ with logarithmic scale in Figure 2.12.

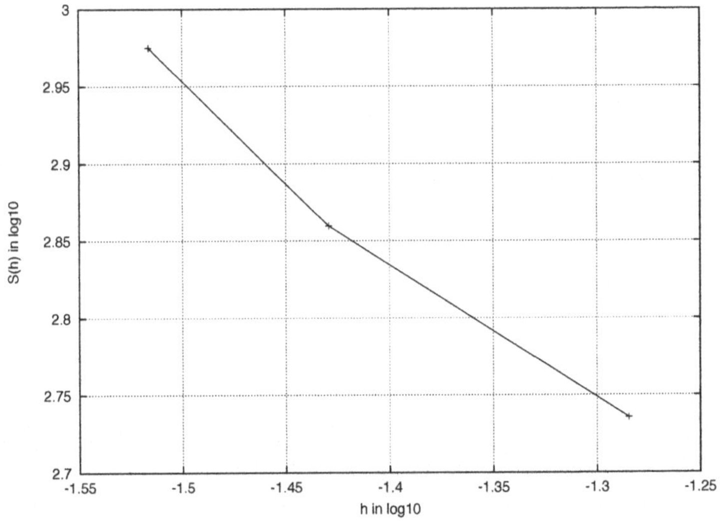

Figure 2.12. *The gradient of the linear regression line is -1.015, which corresponds to the theoretical result $S(h) = C\,h^{-1}$*

3

The ALE Method for Navier-Stokes Equations in a Moving Domain

3.1. Lagrangian and Eulerian coordinates

The formalism used in this section is based on [TEM 01a]. Consider a continuous medium that occupies a domain $\overline{\Omega}_0 \subset \mathbb{R}^2$ at time $t = 0$ and a domain $\overline{\Omega}_t \subset \mathbb{R}^2$ at time $t > 0$. We assume that Ω_0 and Ω_t are open, connected, and bounded sets with Lipschitz boundaries. The word 'domain' does not take its mathematical meaning here. We assume the existence of a map $\varphi : \overline{\Omega}_0 \times [0, T] \rightarrow \mathbb{R}^2$ such that a particle that occupies the initial position $\mathbf{X} \in \overline{\Omega}_0$ will occupy the position $\mathbf{x} = \varphi(\mathbf{X}, t)$ in $\overline{\Omega}_t$. We use the convention $\varphi(\mathbf{X}, t) = \varphi_t(\mathbf{X})$. We call $\mathbf{X} = (X_1, X_2)$ the Lagrangian coordinates and $\mathbf{x} = (x_1, x_2)$ the Eulerian coordinates.

For every $(\mathbf{X}, t) \in \overline{\Omega}_0 \times [0, T]$ we define the displacement and the velocity in Lagrangian coordinates by:

$$\mathbf{U}(\mathbf{X}, t) = \varphi(\mathbf{X}, t) - \mathbf{X},$$

$$\mathbf{V}(\mathbf{X}, t) = \frac{\partial \mathbf{U}}{\partial t}(\mathbf{X}, t)$$

respectively.

We assume that for each $t \in [0, T]$, the map $\varphi_t : \overline{\Omega_0} \to \overline{\Omega_t}$ is bijective. We can define the displacement and the velocity in Eulerian coordinates by:

$$\mathbf{u}(\mathbf{x}, t) = \mathbf{U}\left(\varphi_t^{-1}(\mathbf{x}), t\right),$$

$$\mathbf{v}(\mathbf{x}, t) = \mathbf{V}\left(\varphi_t^{-1}(\mathbf{x}), t\right)$$

respectively, for every $t \in [0, T]$ and $\mathbf{x} \in \overline{\Omega_t}$.

The gradients of a scalar function $q : \Omega_t \to \mathbb{R}$ and a vector-valued function $\mathbf{w} = (w_1, w_2) : \Omega_t \to \mathbb{R}^2$ with respect to Eulerian coordinates are denoted by:

$$\nabla q = \begin{pmatrix} \frac{\partial q}{\partial x_1} \\ \frac{\partial q}{\partial x_2} \end{pmatrix}, \quad \nabla \mathbf{w} = \begin{pmatrix} \frac{\partial w_1}{\partial x_1} & \frac{\partial w_1}{\partial x_2} \\ \frac{\partial w_2}{\partial x_1} & \frac{\partial w_2}{\partial x_2} \end{pmatrix}.$$

The divergence operator acting on a vector-valued function $\mathbf{w} = (w_1, w_2) : \Omega_t \to \mathbb{R}^2$ and on a tensor $\boldsymbol{\sigma} = \left(\sigma_{ij}\right)_{1 \le i,j \le 2}$ with respect to Eulerian coordinates is denoted by:

$$\nabla \cdot \mathbf{w} = \frac{\partial w_1}{\partial x_1} + \frac{\partial w_2}{\partial x_2}, \quad \nabla \cdot \boldsymbol{\sigma} = \begin{pmatrix} \frac{\partial \sigma_{11}}{\partial x_1} + \frac{\partial \sigma_{12}}{\partial x_2} \\ \frac{\partial \sigma_{21}}{\partial x_1} + \frac{\partial \sigma_{22}}{\partial x_2} \end{pmatrix}.$$

Similarly, we use the following notation for the derivatives with respect to the Lagrangian coordinates:

$$\nabla_{\mathbf{X}} \mathbf{U}(\mathbf{X}, t) = \begin{pmatrix} \frac{\partial U_1}{\partial X_1}(\mathbf{X}, t) & \frac{\partial U_1}{\partial X_2}(\mathbf{X}, t) \\ \frac{\partial U_2}{\partial X_1}(\mathbf{X}, t) & \frac{\partial U_2}{\partial X_2}(\mathbf{X}, t) \end{pmatrix},$$

and

$$\nabla_{\mathbf{X}} \cdot \mathbf{V}(\mathbf{X}, t) = \frac{\partial V_1}{\partial X_1}(\mathbf{X}, t) + \frac{\partial V_2}{\partial X_2}(\mathbf{X}, t).$$

If $\mathbf{S} = \left(s_{ij}\right)_{1 \le i,j \le 2}$ is a square matrix, we denote the determinant, the inverse and the transpose by $\det \mathbf{S}$, $\mathbf{S}^{-1}$ and $\mathbf{S}^T$ respectively. We write $\mathbf{S}^{-T} = \left(\mathbf{S}^{-1}\right)^T$, $\operatorname{tr} \mathbf{S} = \sum_{i=1}^{2} s_{ii}$ and $\operatorname{cof} \mathbf{S} = (\det \mathbf{S})\left(\mathbf{S}^{-1}\right)^T$ is the cofactor matrix of $\mathbf{S}$.

We use the following notations: $\mathbf{F}(\mathbf{X}, t) = \mathbf{I} + \nabla_{\mathbf{X}} \mathbf{U}(\mathbf{X}, t)$ for the deformation gradient, $\mathbf{I}$ for the unit (or identity) matrix and $J(\mathbf{X}, t) = \det \mathbf{F}(\mathbf{X}, t)$ for the

Jacobian. We assume that $J(\mathbf{X},t) > 0$ for every $\mathbf{X} \in \Omega_0$ and $t \in [0,T]$. We use $\mathcal{M}_{m,n}(\mathbb{R})$ to denote the set of matrices with coefficients in $\mathbb{R}$ and with m rows and n columns.

PROPOSITION 3.1.– *Consider $S : [0,T] \to \mathcal{M}_{2,2}(\mathbb{R})$ of class $C^1([0,T])$. We assume that for every $t \in [0,T]$, det $(S(t)) \neq 0$. Hence,*

$$\frac{d}{dt}det\,(S(t)) = det\,(S(t))\,tr\left(\frac{dS}{dt}(t)S^{-1}(t)\right).$$

DEMONSTRATION 3.1.– If $S = \left(s_{ij}\right)_{1\leq i,j\leq 2}$, then $\det(S) = s_{11}s_{22} - s_{21}s_{12}$ and

$$\frac{d}{dt}\det(S) = s'_{11}s_{22} + s_{11}s'_{22} - s'_{21}s_{12} - s_{21}s'_{12}.$$

On the other side of the equality, we have

$$\begin{aligned}\operatorname{tr}\left(\frac{dS}{dt}S^{-1}\right) &= \operatorname{tr}\left(\begin{pmatrix} s'_{11} & s'_{12} \\ s'_{21} & s'_{22}\end{pmatrix}\frac{1}{\det(S)}\begin{pmatrix} s_{22} & -s_{12} \\ -s_{21} & s_{11}\end{pmatrix}\right) \\ &= \frac{1}{\det(S)}\operatorname{tr}\left(\begin{pmatrix} s'_{11} & s'_{12} \\ s'_{21} & s'_{22}\end{pmatrix}\begin{pmatrix} s_{22} & -s_{12} \\ -s_{21} & s_{11}\end{pmatrix}\right) \\ &= \frac{1}{\det(S)}\left(s'_{11}s_{22} - s'_{12}s_{21} - s'_{21}s_{12} + s'_{22}s_{11}\right)\end{aligned}$$

giving the desired conclusion. □

PROPOSITION 3.2.– *[Euler's dilatation formula] If φ is of class $C^2(\Omega_0 \times [0,T])$, then*

$$\frac{\partial J}{\partial t}(\mathbf{X},t) = J(\mathbf{X},t)(\nabla\cdot\mathbf{v})(\mathbf{x},t).$$

DEMONSTRATION 3.2.– Since $J(\mathbf{X},t) = \det\mathbf{F}(\mathbf{X},t)$, by applying proposition 3.1 we have

$$\frac{\partial J}{\partial t}(\mathbf{X},t) = J(\mathbf{X},t)\operatorname{tr}\left(\frac{\partial\mathbf{F}}{\partial t}(\mathbf{X},t)\,\mathbf{F}^{-1}(\mathbf{X},t)\right).$$

Since φ is of class C^2, we can change the order of differentiation with respect to time and space

$$\frac{\partial\mathbf{F}}{\partial t}(\mathbf{X},t) = \frac{\partial}{\partial t}(\nabla_{\mathbf{X}}\varphi(\mathbf{X},t)) = \nabla_{\mathbf{X}}\left(\frac{\partial\varphi}{\partial t}(\mathbf{X},t)\right) = \nabla_{\mathbf{X}}\mathbf{V}(\mathbf{X},t).$$

Using the formula for differentiating composite functions, we have

$$\nabla_{\mathbf{X}}\mathbf{V}(\mathbf{X},t) = \nabla_{\mathbf{X}}\mathbf{v}(\varphi(\mathbf{X},t),t) = \nabla\mathbf{v}(\mathbf{x},t)\,\nabla_{\mathbf{X}}\varphi(\mathbf{X},t) = \nabla\mathbf{v}(\mathbf{x},t)\,\mathbf{F}(\mathbf{X},t)$$

and as a consequence

$$\nabla_{\mathbf{X}}\mathbf{V}(\mathbf{X},t)\,\mathbf{F}^{-1}(\mathbf{X},t) = \nabla\mathbf{v}(\mathbf{x},t).$$

Thus,

$$\mathrm{tr}\left(\frac{\partial\mathbf{F}}{\partial t}(\mathbf{X},t)\,\mathbf{F}^{-1}(\mathbf{X},t)\right) = \mathrm{tr}\,(\nabla\mathbf{v}(\mathbf{x},t)) = (\nabla\cdot\mathbf{v})(\mathbf{x},t)$$

giving the desired conclusion. □

It is not necessary to require $\varphi \in C^2$, it is enough that $\frac{\partial^2\varphi}{\partial X_i\partial t}, \frac{\partial^2\varphi}{\partial t\partial X_i}$ exist and

$$\frac{\partial^2\varphi}{\partial X_i\partial t} = \frac{\partial^2\varphi}{\partial t\partial X_i}.$$

Let $H : \overline{\Omega}_0 \times [0,T] \to \mathbb{R}$. We define for each $t \in [0,T]$ and $\mathbf{x} \in \overline{\Omega}_t$:

$$h(\mathbf{x},t) = H\left(\varphi_t^{-1}(\mathbf{x}),t\right).$$

We define the *material derivative* by

$$\frac{Dh}{Dt}(\mathbf{x},t) = \frac{\partial H}{\partial t}(\mathbf{X},t).$$

We have

$$
\begin{aligned}
\frac{\partial H}{\partial t}(\mathbf{X},t) &= \frac{d}{dt}h(\varphi(\mathbf{X},t),t) \\
&= \frac{\partial h}{\partial x_1}(\varphi(\mathbf{X},t),t)\frac{\partial \varphi_1}{\partial t}(\mathbf{X},t) + \frac{\partial h}{\partial x_2}(\varphi(\mathbf{X},t),t)\frac{\partial \varphi_2}{\partial t}(\mathbf{X},t) \\
&\quad + \frac{\partial h}{\partial t}(\varphi(\mathbf{X},t),t) \\
&= \frac{\partial h}{\partial x_1}(\mathbf{x},t)\,V_1(\mathbf{X},t) + \frac{\partial h}{\partial x_2}(\mathbf{x},t)\,V_2(\mathbf{X},t) + \frac{\partial h}{\partial t}(\mathbf{x},t) \\
&= \frac{\partial h}{\partial x_1}(\mathbf{x},t)\,v_1(\mathbf{x},t) + \frac{\partial h}{\partial x_2}(\mathbf{x},t)\,v_2(\mathbf{x},t) + \frac{\partial h}{\partial t}(\mathbf{x},t) \\
&= \mathbf{v}(\mathbf{x},t)\cdot\nabla h(\mathbf{x},t) + \frac{\partial h}{\partial t}(\mathbf{x},t). \qquad [3.1]
\end{aligned}
$$

PROPOSITION 3.3.– *[Reynolds transport formula] If H is of class $C^1(\Omega_0 \times [0,T])$ and φ is of class $C^2(\Omega_0 \times [0,T])$, then:*

$$
\begin{aligned}
\frac{d}{dt}\int_{\Omega_t} h(\mathbf{x},t)\,d\mathbf{x} &= \int_{\Omega_t}\left(\frac{Dh}{Dt}(\mathbf{x},t) + h(\mathbf{x},t)(\nabla\cdot\mathbf{v})(\mathbf{x},t)\right)d\mathbf{x} \\
&= \int_{\Omega_t}\left(\frac{\partial h}{\partial t}(\mathbf{x},t) + (\nabla\cdot(h\mathbf{v}))(\mathbf{x},t)\right)d\mathbf{x}.
\end{aligned}
$$

DEMONSTRATION 3.3.– We have

$$
\int_{\Omega_t} h(\mathbf{x},t)\,d\mathbf{x} = \int_{\Omega_0} h(\varphi(\mathbf{X},t),t)\,J(\mathbf{X},t)\,d\mathbf{X} = \int_{\Omega_0} H(\mathbf{X},t)\,J(\mathbf{X},t)\,d\mathbf{X}
$$

giving

$$
\frac{d}{dt}\int_{\Omega_t} h(\mathbf{x},t)\,d\mathbf{x} = \frac{d}{dt}\int_{\Omega_0} H(\mathbf{X},t)\,J(\mathbf{X},t)\,d\mathbf{X} = \int_{\Omega_0}\left(\frac{\partial H}{\partial t}J + H\frac{\partial J}{\partial t}\right)d\mathbf{X}.
$$

Applying the definition of the material derivative and proposition 3.2, we obtain the following equality:

$$\int_{\Omega_0}\left(\frac{\partial H}{\partial t}J + H\frac{\partial J}{\partial t}\right)d\mathbf{X} = \int_{\Omega_0}\frac{Dh}{Dt}\left(\varphi\left(\mathbf{X},t\right),t\right)J\left(\mathbf{X},t\right)d\mathbf{X}$$

$$+\int_{\Omega_0}H\left(\mathbf{X},t\right)J\left(\mathbf{X},t\right)\left(\nabla\cdot\mathbf{v}\right)\left(\varphi\left(\mathbf{X},t\right),t\right)d\mathbf{X}$$

$$=\int_{\Omega_t}\frac{Dh}{Dt}\left(\mathbf{x},t\right)d\mathbf{x}+\int_{\Omega_t}h\left(\mathbf{x},t\right)\left(\nabla\cdot\mathbf{v}\right)\left(\mathbf{x},t\right)d\mathbf{x}.$$

Finally, if we use [3.1], we have

$$\int_{\Omega_t}\frac{Dh}{Dt}\left(\mathbf{x},t\right)d\mathbf{x}+\int_{\Omega_t}h\left(\mathbf{x},t\right)\left(\nabla\cdot\mathbf{v}\right)\left(\mathbf{x},t\right)d\mathbf{x}$$

$$=\int_{\Omega_t}\frac{\partial h}{\partial t}\left(\mathbf{x},t\right)d\mathbf{x}+\int_{\Omega_t}\mathbf{v}\left(\mathbf{x},t\right)\cdot\nabla h\left(\mathbf{x},t\right)d\mathbf{x}+\int_{\Omega_t}h\left(\mathbf{x},t\right)\left(\nabla\cdot\mathbf{v}\right)\left(\mathbf{x},t\right)d\mathbf{x}$$

$$=\int_{\Omega_t}\frac{\partial h}{\partial t}\left(\mathbf{x},t\right)d\mathbf{x}+\int_{\Omega_t}\left(\nabla\cdot\left(h\mathbf{v}\right)\right)\left(\mathbf{x},t\right)d\mathbf{x}.$$

□

3.2. Arbitrary Lagrangian Eulerian (ALE) coordinates

Consider a fluid that occupies a domain Ω_0 in $\mathbb{R}^2$ at time $t = 0$ and a domain Ω_t at time $t > 0$. In fluid mechanics, we generally use Eulerian coordinates.

Let $h(\cdot, t) : \Omega_t \to \mathbb{R}$. To approximate the time derivative at the point $\mathbf{x} \in \Omega_t$, we can use the scheme

$$\frac{\partial h}{\partial t}\left(\mathbf{x},t\right)\approx\frac{h\left(\mathbf{x},t\right)-h\left(\mathbf{x},t-\Delta t\right)}{\Delta t}$$

but, because the fluid domain is moving, the point $\mathbf{x}$ does not necessarily belong in $\Omega_{t-\Delta t}$, thus $h(\mathbf{x}, t - \Delta t)$ is not well-defined.

We can change variables to stay within a fixed domain. If we have the velocities $\mathbf{v}$ in Eulerian coordinates, we can find the trajectory, $\mathbf{x}(t)$, of a particle that occupies the position $\mathbf{X}$ at time $t = 0$ as the solution of

$$\frac{d\mathbf{x}}{dt}(t) = \mathbf{v}(\mathbf{x}(t), t)$$
$$\mathbf{x}(0) = \mathbf{X}.$$

We define $\varphi_t(\mathbf{X}) = \mathbf{x}(t)$ and we can use Lagrangian coordinates to write the variables in the fixed domain Ω_0. However, we can do things more simply.

We use $\widehat{\Omega}$ to denote a fixed reference domain, which is an open, connected, bounded set in $\mathbb{R}^2$, with Lipschitz boundary. Let $\mathcal{A}_t : \overline{\widehat{\Omega}} \to \overline{\Omega}_t$, $t \in [0, T]$ be a family of bijective maps. We call $\widehat{\mathbf{x}} = (\widehat{x}_1, \widehat{x}_2) \in \widehat{\Omega}$ the ALE coordinates and $\mathbf{x} = (x_1, x_2) = \mathcal{A}_t(\widehat{\mathbf{x}})$ are the Eulerian coordinates. We assume that $\mathcal{A}_t$ and the inverse $\mathcal{A}_t^{-1}$ are of class C^1. We also use the convention $\mathcal{A}(\widehat{\mathbf{x}}, t) = \mathcal{A}_t(\widehat{\mathbf{x}})$.

In what follows, we denote the Jacobian matrix of the ALE map, $\widehat{\mathbf{x}} \to \mathcal{A}_t(\widehat{\mathbf{x}})$, with $\nabla_{\widehat{\mathbf{x}}}\mathcal{A}_t(\widehat{\mathbf{x}})$ and the Jacobian with $\widehat{J}(\widehat{\mathbf{x}}, t) = \det(\nabla_{\widehat{\mathbf{x}}}\mathcal{A}_t(\widehat{\mathbf{x}}))$. We assume that the derivative $\frac{\partial \mathcal{A}_t}{\partial t}(\widehat{\mathbf{x}})$ exists for every $\mathbf{x} \in \widehat{\Omega}$ and we denote the *domain velocity* with

$$\boldsymbol{\vartheta}(\mathbf{x}, t) = \frac{\partial \mathcal{A}}{\partial t}(\widehat{\mathbf{x}}, t).$$

Let $\mathbf{v}$ be the fluid velocity in Eulerian coordinates. We shall define the velocity in ALE coordinates, $\widehat{\mathbf{v}} : \widehat{\Omega} \times [0, T] \to \mathbb{R}^2$, by

$$\widehat{\mathbf{v}}(\widehat{\mathbf{x}}, t) = \mathbf{v}(\mathcal{A}_t(\widehat{\mathbf{x}}), t) = \mathbf{v}(\mathbf{x}, t).$$

Similar to the material derivative, we can define the *ALE time derivative* by:

$$\left.\frac{\partial \mathbf{v}}{\partial t}\right|_{\widehat{\mathbf{x}}}(\mathbf{x}, t) = \frac{\partial \widehat{\mathbf{v}}}{\partial t}(\widehat{\mathbf{x}}, t).$$

In the same way that we obtained [3.1], by replacing Ω_0 with $\widehat{\Omega}$, φ with $\mathcal{A}$ and $\mathbf{X}$ with $\widehat{\mathbf{x}}$, we can deduce

$$\left.\frac{\partial v_1}{\partial t}\right|_{\widehat{\mathbf{x}}}(\mathbf{x}, t) = \frac{\partial v_1}{\partial t}(\mathbf{x}, t) + \boldsymbol{\vartheta}(\mathbf{x}, t) \cdot \nabla v_1(\mathbf{x}, t)$$
$$\left.\frac{\partial v_2}{\partial t}\right|_{\widehat{\mathbf{x}}}(\mathbf{x}, t) = \frac{\partial v_2}{\partial t}(\mathbf{x}, t) + \boldsymbol{\vartheta}(\mathbf{x}, t) \cdot \nabla v_2(\mathbf{x}, t)$$

or equivalently

$$\left.\frac{\partial \mathbf{v}}{\partial t}\right|_{\widehat{\mathbf{x}}}(\mathbf{x},t) = \frac{\partial \mathbf{v}}{\partial t}(\mathbf{x},t) + (\boldsymbol{\vartheta}\cdot\nabla)\,\mathbf{v}(\mathbf{x},t). \qquad [3.2]$$

PROPOSITION 3.4.– *[the ALE dilatation formula] If $\mathcal{A}$ is of class $C^2\left(\widehat{\Omega}\times[0,T]\right)$, then*

$$\frac{\partial \widehat{J}}{\partial t}(\widehat{\mathbf{x}},t) = \widehat{J}(\widehat{\mathbf{x}},t)\,(\nabla\cdot\boldsymbol{\vartheta})(\mathbf{x},t).$$

Let $h:(\cdot,t):\Omega_t \to \mathbb{R}$. We define for every $t \in [0,T]$ and $\widehat{\mathbf{x}} \in \widehat{\Omega}$:

$$\widehat{h}(\widehat{\mathbf{x}},t) = h(\mathcal{A}_t(\widehat{\mathbf{x}}),t).$$

PROPOSITION 3.5.– *[the ALE transport formula] If $\widehat{h}$ is of class $C^1\left(\widehat{\Omega}\times[0,T]\right)$ and $\mathcal{A}$ is of class $C^2\left(\widehat{\Omega}\times[0,T]\right)$, then:*

$$\begin{aligned}\frac{d}{dt}\int_{\Omega_t} h(\mathbf{x},t)\,d\mathbf{x} &= \int_{\Omega_t}\left(\left.\frac{\partial h}{\partial t}\right|_{\widehat{\mathbf{x}}}(\mathbf{x},t) + h(\mathbf{x},t)\,(\nabla\cdot\boldsymbol{\vartheta})(\mathbf{x},t)\right)d\mathbf{x}\\ &= \int_{\Omega_t}\left(\frac{\partial h}{\partial t} + \boldsymbol{\vartheta}\cdot\nabla h + (\nabla\cdot\boldsymbol{\vartheta})\,h\right)d\mathbf{x}\\ &= \int_{\Omega_t}\left(\frac{\partial h}{\partial t}(\mathbf{x},t) + (\nabla\cdot(h\boldsymbol{\vartheta}))(\mathbf{x},t)\right)d\mathbf{x}.\end{aligned}$$

3.2.1. *Construction of an ALE map by harmonic extension*

Let Ω_t and $\widehat{\Omega}$ be two domains in $\mathbb{R}^2$. We assume that there exists a map $\widehat{\mathbf{u}}(\cdot,t):\partial\widehat{\Omega} \to \mathbb{R}^2$ such that

$$\widehat{\mathbf{x}} \to \widehat{\mathbf{x}} + \widehat{\mathbf{u}}(\widehat{\mathbf{x}},t)$$

is a bijection from $\partial\widehat{\Omega}$ to $\partial\Omega_t$. We can define an ALE map by:

$$\mathcal{A}_t : \overline{\widehat{\Omega}} \to \mathbb{R}^2, \quad \mathcal{A}_t(\widehat{\mathbf{x}}) = \widehat{\mathbf{x}} + \widehat{\mathbf{d}}(\widehat{\mathbf{x}},t)$$

where $\widehat{\mathbf{d}}(\cdot,t) : \overline{\widehat{\Omega}} \to \mathbb{R}^2$ is the solution of the Dirichlet problem

$$\Delta\widehat{\mathbf{d}}(\cdot,t) = 0, \text{in } \widehat{\Omega}, \quad \widehat{\mathbf{d}}(\cdot,t) = \widehat{\mathbf{u}}(\cdot,t), \text{on } \partial\widehat{\Omega}. \qquad [3.3]$$

REMARK 3.1.– *Using [BAR 98], Theorem 2.8.1, p. 78, if $\widehat{\Omega}$ is of class C^2 and $\widehat{\mathbf{u}}(\cdot,t) \in C^0\left(\partial\widehat{\Omega}\right)$, then $\widehat{\mathbf{d}}(\cdot,t) \in C^2\left(\widehat{\Omega}\right) \cap C^0\left(\overline{\widehat{\Omega}}\right)$. We have a stronger result ([BAR 98], Theorem 2.7.1, p. 71) that if every point of $\partial\widehat{\Omega}$ is regular and $\widehat{\mathbf{u}}(\cdot,t) \in C^0\left(\partial\widehat{\Omega}\right)$, then we obtain the same conclusion. A sufficient condition in $\mathbb{R}^d$, $d = 2, 3$ for every point on the boundary to be regular is for the domain to satisfy the exterior cone property, i.e. for every $x_0 \in \partial\widehat{\Omega}$, there exists an open cone C and $r_0 > 0$, such that*

$$\widehat{\Omega} \cap \{C + x_0\} \cap B(x_0, r_0) = \emptyset$$

where $B(x_0, r_0)$ is the open ball with center x_0 and radius r_0. If the domain is bounded and its boundary is Lipschitz, then it has the exterior cone property (see [HEN 05] Theorem 2.4.7, p. 53 and Remark 2.4.9, p. 56).

3.2.2. *Construction of an explicit ALE map*

Let L, H be two positive constants. For every $t \in [0,T]$, let $\widehat{u}_t : [0,L] \to \mathbb{R}$ be the function in $C^0([0,L]) \cap C^1(]0,L[)$ such that $\widehat{u}_t(0) = \widehat{u}_t(L) = 0$ and $H + \min_{x_1 \in [0,L]} \widehat{u}_t(x_1) > 0$. We shall introduce the domain (see Figure 3.1)

$$\Omega_t = \left\{(x_1,x_2) \in \mathbb{R}^2;\ x_1 \in]0,L[,\ 0 < x_2 < H + \widehat{u}_t(x_1)\right\}$$

which has the boundaries

$$\Gamma_{in} = \left\{(0,x_2) \in \mathbb{R}^2;\ x_2 \in]0,H[\right\}$$

$$\Gamma_D = \left\{(x_1,0) \in \mathbb{R}^2;\ x_1 \in]0,L[\right\}$$

$$\Gamma_{out} = \left\{(L,x_2) \in \mathbb{R}^2;\ x_2 \in]0,H[\right\}$$

$$\Gamma_t = \left\{(x_1,x_2) \in \mathbb{R}^2;\ x_1 \in]0,L[,\ x_2 = H + \widehat{u}_t(x_1)\right\}.$$

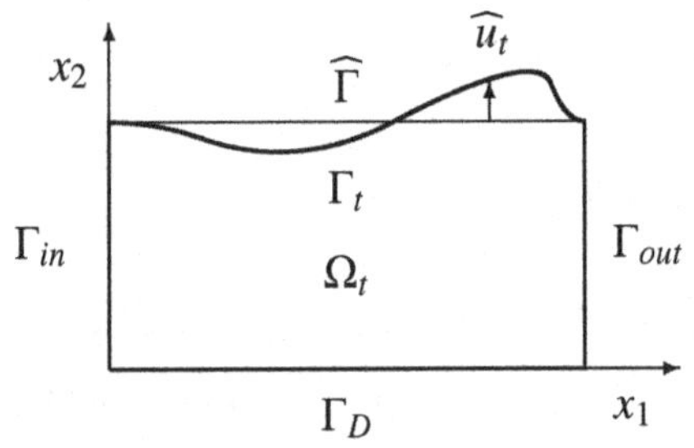

Figure 3.1. *The fluid domain with moving upper boundary*

We denote the ALE reference domain and its upper boundary by:

$$\widehat{\Omega} =]0, L[\times]0, H[, \quad \widehat{\Gamma} = \left\{(x_1, H) \in \mathbb{R}^2;\ x_1 \in]0, L[\right\}.$$

We can define the ALE map, $\mathcal{A}_t : \overline{\widehat{\Omega}} \to \overline{\Omega}_t$ by:

$$\mathcal{A}_t\left(\widehat{x_1}, \widehat{x_2}\right) = \left(\widehat{x_1}, \frac{H + \widehat{u}_t\left(\widehat{x_1}\right)}{H}\widehat{x_2}\right)$$

which has the inverse

$$\mathcal{A}_t^{-1}\left(x_1, x_2\right) = \left(x_1, \frac{Hx_2}{H + \widehat{u}_t\left(x_1\right)}\right)$$

and which satisfies $\mathcal{A}_t\left(\widehat{\Omega}\right) = \Omega_t$, $\mathcal{A}_t\left(\widehat{\Gamma}\right) = \Gamma_t$ and $\mathcal{A}_t\left(\widehat{x}\right) = \widehat{x}$ for every $\widehat{x} \in \Gamma_{in} \cup \Gamma_D \cup \Gamma_{out}$.

If $\mathbf{v}(\cdot, t) : \Omega_t \to \mathbb{R}^2$ and $p(\cdot, t) : \Omega_t \to \mathbb{R}$ are the fluid velocity and pressure in Eulerian coordinates, we shall use $\widehat{\mathbf{v}} : \widehat{\Omega} \times [0, T] \to \mathbb{R}^2$ and $\widehat{p} : \widehat{\Omega} \times [0, T] \to \mathbb{R}$ to denote the fluid velocity and pressure in ALE coordinates defined by $\widehat{\mathbf{v}}(\widehat{\mathbf{x}}, t) = \mathbf{v}\left(\mathcal{A}_t(\widehat{\mathbf{x}}), t\right)$ and $\widehat{p}(\widehat{\mathbf{x}}, t) = p\left(\mathcal{A}_t(\widehat{\mathbf{x}}), t\right)$.

Using the formula for differentiating a composite function, we have

$$\nabla_{\widehat{\mathbf{x}}}\widehat{\mathbf{v}}(\widehat{\mathbf{x}}) = \nabla\mathbf{v}(\mathbf{x})\nabla_{\widehat{\mathbf{x}}}\mathcal{A}_t(\widehat{\mathbf{x}})$$

which implies

$$\nabla \mathbf{v}(\mathbf{x}) = \nabla_{\widehat{\mathbf{x}}} \widehat{\mathbf{v}}(\widehat{\mathbf{x}}) \left(\nabla_{\widehat{\mathbf{x}}} \mathcal{A}_t(\widehat{\mathbf{x}}) \right)^{-1}.$$

We have

$$\nabla_{\widehat{\mathbf{x}}} \mathcal{A}_t(\widehat{\mathbf{x}}) = \begin{pmatrix} 1 & 0 \\ \frac{\partial \widehat{u}_t}{\partial \widehat{x}_1}(\widehat{x}_1) \frac{\widehat{x}_2}{H} & \frac{H+\widehat{u}_t(\widehat{x}_1)}{H} \end{pmatrix}$$

giving $\widehat{J}(\widehat{\mathbf{x}}, t) = \det(\nabla_{\widehat{\mathbf{x}}} \mathcal{A}_t(\widehat{\mathbf{x}})) = \frac{H+\widehat{u}_t(\widehat{x}_1)}{H}$. We can obtain the following relation:

$$(\nabla_{\widehat{\mathbf{x}}} \mathcal{A}_t(\widehat{\mathbf{x}}))^{-1} = \begin{pmatrix} 1 & 0 \\ -\frac{\partial \widehat{u}_t}{\partial \widehat{x}_1}(\widehat{x}_1) \frac{\widehat{x}_2}{H+\widehat{u}_t(\widehat{x}_1)} & \frac{H}{H+\widehat{u}_t(\widehat{x}_1)} \end{pmatrix}.$$

We observe that $\mathcal{A}_t \in C^1(\widehat{\Omega})$, $\mathcal{A}_t^{-1} \in C^1(\Omega_t)$ and we say that $\mathcal{A}_t$ is a C^1 diffeomorphism.

The incompressibility condition of a fluid in Eulerian coordinates is given as follows:

$$(\nabla \cdot \mathbf{v})(\mathbf{x}, t) = \frac{\partial v_1}{\partial x_1}(\mathbf{x}, t) + \frac{\partial v_2}{\partial x_2}(\mathbf{x}, t) = 0$$

and using ALE coordinates is given as follows:

$$\frac{\partial \widehat{v}_1}{\partial \widehat{x}_1}(\widehat{\mathbf{x}}, t) - \left(\frac{\partial \widehat{u}_t}{\partial \widehat{x}_1}(\widehat{x}_1) \frac{\widehat{x}_2}{H + \widehat{u}_t(\widehat{x}_1)} \right) \frac{\partial \widehat{v}_1}{\partial \widehat{x}_2}(\widehat{\mathbf{x}}, t) + \left(\frac{H}{H + \widehat{u}_t(\widehat{x}_1)} \right) \frac{\partial \widehat{v}_2}{\partial \widehat{x}_2}(\widehat{\mathbf{x}}, t) = 0$$

which is more complicated!

3.3. Weak non-conservative and conservative formulations

We shall study the Navier-Stokes equations in a moving domain $\overline{\Omega}_t$, $t \in [0, T]$, as in Figure 3.1. We consider the case where the motion of domain Ω_t is known, and so we are not dealing with a free boundary problem where the domain motion is one of the unknowns in the problem.

At each time $t \in [0,T]$, we are looking for the velocity $\mathbf{v}(\cdot,t) = (v_1(\cdot,t), v_2(\cdot,t)) : \overline{\Omega_t} \to \mathbb{R}^2$ and the pressure $p(\cdot,t) : \overline{\Omega_t} \to \mathbb{R}$, such that

$$\rho\left(\frac{\partial \mathbf{v}}{\partial t} + (\mathbf{v}\cdot\nabla)\mathbf{v}\right) - 2\mu\nabla\cdot\epsilon(\mathbf{v}) + \nabla p = \mathbf{f},\ \forall t \in]0,T[, \forall \mathbf{x} \in \Omega_t \quad [3.4]$$

$$\nabla\cdot\mathbf{v} = 0,\ \forall t \in]0,T[, \forall \mathbf{x} \in \Omega_t \quad [3.5]$$

$$\sigma\mathbf{n} = \mathbf{h}_{in},\ \text{on } \Gamma_{in}\times]0,T[\quad [3.6]$$

$$\sigma\mathbf{n} = \mathbf{h}_{out},\ \text{on } \Gamma_{out}\times]0,T[\quad [3.7]$$

$$\mathbf{v} = 0,\ \text{on } \Gamma_D\times]0,T[\quad [3.8]$$

$$\mathbf{v} = \mathbf{g},\ \forall t \in]0,T[, \forall \mathbf{x} \in \Gamma_t \quad [3.9]$$

$$\mathbf{v}(\mathbf{x},0) = \mathbf{v}_0(\mathbf{x}),\ \forall \mathbf{x} \in \Omega_0 \quad [3.10]$$

where

– ρ, the mass density, and μ, the viscosity, are two positive constants,

– $\mathbf{f}(\cdot,t) = (f_1(\cdot,t), f_2(\cdot,t)) : \Omega_t \to \mathbb{R}^2$ are the body forces, for example, $f_1 = 0$ and $f_2 = -\rho\, 9.81$ for the gravitational forces,

– $\mathbf{h}_{in} : \Gamma_{in}\times]0,T[\to \mathbb{R}^2$ and $\mathbf{h}_{out} : \Gamma_{out}\times]0,T[\to \mathbb{R}^2$ are the forces at the entrance and the exit,

– $\mathbf{g}(\cdot,t) : \Gamma_t \to \mathbb{R}^2$ is the velocity imposed on the moving boundary,

– $\mathbf{v}_0 : \Omega_0 \to \mathbb{R}^2$ is the initial fluid velocity.

We have used the notation $\epsilon(\mathbf{v}) = \frac{1}{2}\left(\nabla\mathbf{v} + (\nabla\mathbf{v})^T\right)$ for the strain rate tensor, $\sigma = -p\mathbf{I} + 2\mu\epsilon\left(\mathbf{v}^F\right)$ for the stress tensor, $\mathbf{n}$ for the external unit vector normal to the boundary and $\mathbf{I}$ for the unit matrix.

We denote the ALE reference domain and its upper boundary with

$$\widehat{\Omega} =]0,L[\times]0,H[, \quad \widehat{\Gamma} = \left\{(x_1,H) \in \mathbb{R}^2;\ x_1 \in]0,L[\right\}$$

and let $\mathcal{A}_t : \overline{\widehat{\Omega}} \to \overline{\Omega_t}$ be an ALE map such that

$$\mathcal{A}_t(\widehat{\Omega}) = \Omega_t, \quad \mathcal{A}_t(\widehat{\Gamma}) = \Gamma_t, \quad \mathcal{A}_t(\widehat{\mathbf{x}}) = \widehat{\mathbf{x}}, \quad \forall \widehat{\mathbf{x}} \in \Gamma_{in} \cup \Gamma_D \cup \Gamma_{out}.$$

If $\widehat{\mathbf{v}} : \widehat{\Omega} \times [0,T] \to \mathbb{R}^2$, we can define for every $t \in [0,T]$ and $\mathbf{x} \in \Omega_t$

$$\mathbf{v}(\mathbf{x},t) = \widehat{\mathbf{v}}\left(\mathcal{A}_t^{-1}(\mathbf{x}), t\right).$$

Starting from a space of functions in ALE coordinates, for example $C^0\left([0,T], L^2(\widehat{\Omega})\right)$, we can construct the associated space of functions in Eulerian coordinates. If $\widehat{p} \in C^0\left([0,T], L^2(\widehat{\Omega})\right)$, then $\widehat{p}(t) \in L^2(\widehat{\Omega})$. We can write $\widehat{p}(\widehat{x},t)$ instead of $\widehat{p}(t)(\widehat{x})$. If $\mathcal{A}_t$ is a C^1 diffeomorphism, *i.e.* $\mathcal{A}_t \in C^1(\widehat{\Omega})$ and $\mathcal{A}_t^{-1} \in C^1(\Omega_t)$, then $p(\cdot,t) \in L^2(\Omega_t)$. Analogously, if $\widehat{\mathbf{v}} \in C^0([0,T], (H^1(\widehat{\Omega}))^2)$, then $\mathbf{v}(\cdot,t) \in (H^1(\Omega_t))^2$.

Using [3.2], we have

$$\frac{\partial \mathbf{v}}{\partial t} + (\mathbf{v}\cdot\nabla)\mathbf{v} = \left.\frac{\partial \mathbf{v}}{\partial t}\right|_{\widehat{\mathbf{x}}} - (\boldsymbol{\vartheta}\cdot\nabla)\,\mathbf{v} + (\mathbf{v}\cdot\nabla)\mathbf{v} = \left.\frac{\partial \mathbf{v}}{\partial t}\right|_{\widehat{\mathbf{x}}} + ((\mathbf{v}-\boldsymbol{\vartheta})\cdot\nabla)\,\mathbf{v}.$$

3.3.1. *Weak non-conservative formulation*

Find $\mathbf{v}$ and p, such that $\widehat{\mathbf{v}} \in C^1\left(]0,T[, (L^2(\widehat{\Omega}))^2\right) \cap C^0\left([0,T], (H^1(\widehat{\Omega}))^2\right)$, $\widehat{p} \in C^0\left([0,T], L^2(\widehat{\Omega})\right)$, $\mathbf{v} = \mathbf{g}$ on Γ_t, $\mathbf{v} = 0$ on Γ_D and

$$\int_{\Omega_t} \rho \left.\frac{\partial \mathbf{v}}{\partial t}\right|_{\widehat{\mathbf{x}}} \cdot \mathbf{w}\, d\mathbf{x} + \int_{\Omega_t} \rho[((\mathbf{v}-\boldsymbol{\vartheta})\cdot\nabla)\mathbf{v}]\cdot\mathbf{w}\, d\mathbf{x}$$

$$+ \int_{\Omega_t} 2\mu\, \boldsymbol{\epsilon}(\mathbf{v}) : \boldsymbol{\epsilon}(\mathbf{w})\, d\mathbf{x} - \int_{\Omega_t} (\nabla\cdot\mathbf{w})\, p\, d\mathbf{x} = \int_{\Omega_t} \mathbf{f}\cdot\mathbf{w}\, d\mathbf{x} + \int_{\Gamma_N} \mathbf{h}\cdot\mathbf{w}\, ds,$$

$$\forall \mathbf{w} \in (H^1(\Omega_t))^2,\ \mathbf{w} = 0 \text{ on } \Gamma_D \cup \Gamma_t, \qquad [3.11]$$

$$-\int_{\Omega_t} (\nabla\cdot\mathbf{v})\, q\, d\mathbf{x} = 0, \quad \forall q \in L^2(\Omega_t), \qquad [3.12]$$

$$\mathbf{v}(\cdot,0) = \mathbf{v}_0(\cdot). \qquad [3.13]$$

We have introduced $\Gamma_N = \Gamma_{in} \cup \Gamma_{out}$ and $\mathbf{h} : \Gamma_N \times [0,T] \to \mathbb{R}^2$, defined by $\mathbf{h} = \mathbf{h}_{in}$ on Γ_{in} and $\mathbf{h} = \mathbf{h}_{out}$ on Γ_{out}. With regards to the regularity of the values, we assume that $\mathbf{v}_0 \in (H^1(\Omega_0))^2$ and

$$\mathbf{h} \in C^0\left([0,T], (L^2(\Gamma_N))^2\right),$$
$$\widehat{\mathbf{f}} \in C^0\left([0,T], (L^2(\widehat{\Omega}))^2\right),$$
$$\widehat{\mathbf{g}} \in C^0\left([0,T], (H^{1/2}(\widehat{\Gamma}))^2\right).$$

We can generate test functions on Ω_t from test functions on $\widehat{\Omega}$.

The spaces of test functions in the ALE domain are as follows:

$$\widehat{W} = \left\{\widehat{\mathbf{w}} \in \left(H^1\left(\widehat{\Omega}\right)\right)^2;\ \widehat{\mathbf{w}} = 0 \text{ on } \Gamma_D \cup \widehat{\Gamma}\right\}$$
$$\widehat{Q} = L^2\left(\widehat{\Omega}\right).$$

We shall introduce the spaces of test functions in Eulerian coordinates

$$W = \Big\{\mathbf{w}(\cdot,t) : \Omega_t \to \mathbb{R}^2;$$
$$\exists \widehat{\mathbf{w}} \in \widehat{W},\ \forall t \in [0,T],\ \forall \mathbf{x} \in \Omega_t,\ \mathbf{w}(\mathbf{x},t) = \widehat{\mathbf{w}}\left(\mathcal{A}_t^{-1}(\mathbf{x})\right)\Big\}$$
$$Q = \Big\{q(\cdot,t) : \Omega_t \to \mathbb{R};\ \exists \widehat{q} \in \widehat{Q},\ \forall t \in [0,T],$$
$$\forall \mathbf{x} \in \Omega_t,\ q(\mathbf{x},t) = \widehat{q}\left(\mathcal{A}_t^{-1}(\mathbf{x})\right)\Big\}.$$

It should be noted that $\widehat{\mathbf{w}} \in \widehat{W}$ does not depend on time, but $\mathbf{w} \in W$ does!

Using the ALE transport formula, we have

$$\frac{d}{dt}\int_{\Omega_t} \mathbf{v}\cdot\mathbf{w}\,d\mathbf{x} = \int_{\Omega_t} \left.\frac{\partial(\mathbf{v}\cdot\mathbf{w})}{\partial t}\right|_{\widehat{\mathbf{x}}} d\mathbf{x} + \int_{\Omega_t} (\nabla\cdot\boldsymbol{\vartheta})(\mathbf{v}\cdot\mathbf{w})\,d\mathbf{x}$$
$$= \int_{\Omega_t} \left.\frac{\partial \mathbf{v}}{\partial t}\right|_{\widehat{\mathbf{x}}} \cdot \mathbf{w}\,d\mathbf{x} + \int_{\Omega_t} \mathbf{v}\cdot \left.\frac{\partial \mathbf{w}}{\partial t}\right|_{\widehat{\mathbf{x}}} d\mathbf{x} + \int_{\Omega_t} (\nabla\cdot\boldsymbol{\vartheta})(\mathbf{v}\cdot\mathbf{w})\,d\mathbf{x}.$$

However,

$$\left.\frac{\partial \mathbf{w}}{\partial t}\right|_{\widehat{\mathbf{x}}}(\mathbf{x},t) = \frac{\partial \widehat{\mathbf{w}}}{\partial t}(\widehat{\mathbf{x}}) = 0$$

because the test function does not depend on time in ALE coordinates, thus

$$\int_{\Omega_t} \left.\frac{\partial \mathbf{v}}{\partial t}\right|_{\widehat{\mathbf{x}}} \cdot \mathbf{w}\, d\mathbf{x} = \frac{d}{dt}\int_{\Omega_t} \mathbf{v}\cdot\mathbf{w}\, d\mathbf{x} - \int_{\Omega_t} (\nabla\cdot\boldsymbol{\vartheta})(\mathbf{v}\cdot\mathbf{w})\, d\mathbf{x}.$$

Substituting this equation into [3.11], we obtain an equivalent formulation.

3.3.2. *Weak conservative formulation*

Find $\mathbf{v}$ and p, such that $\mathbf{v} = \mathbf{g}$ on Γ_t, $\mathbf{v} = 0$ on Γ_D and $\forall \mathbf{w} \in W$ and $\forall q \in Q$ we have

$$\rho\frac{d}{dt}\int_{\Omega_t} \mathbf{v}\cdot\mathbf{w}\, d\mathbf{x} - \int_{\Omega_t} \rho(\nabla\cdot\boldsymbol{\vartheta})(\mathbf{v}\cdot\mathbf{w})\, d\mathbf{x} + \int_{\Omega_t} \rho[((\mathbf{v}-\boldsymbol{\vartheta})\cdot\nabla)\mathbf{v}]\cdot\mathbf{w}\, d\mathbf{x}$$

$$+\int_{\Omega_t} 2\mu\,\boldsymbol{\epsilon}(\mathbf{v}) : \boldsymbol{\epsilon}(\mathbf{w})\, d\mathbf{x} - \int_{\Omega_t} (\nabla\cdot\mathbf{w})\, p\, d\mathbf{x}$$

$$= \int_{\Omega_t} \mathbf{f}\cdot\mathbf{w}\, d\mathbf{x} + \int_{\Gamma_N} \mathbf{h}\cdot\mathbf{w}\, ds, \qquad [3.14]$$

$$-\int_{\Omega_t} (\nabla\cdot\mathbf{v})\, q\, d\mathbf{x} = 0, \qquad [3.15]$$

$$\mathbf{v}(\cdot, 0) = \mathbf{v}_0(\cdot). \qquad [3.16]$$

We use the same regularity as for the non-conservative formulation.

We define

$$\mathbf{v}\otimes\mathbf{w} = (v_i w_j)_{1\le i,j\le 2},$$

hence,

$$\nabla\cdot(\mathbf{v}\otimes\mathbf{w}) = (\mathbf{w}\cdot\nabla)\mathbf{v} + (\nabla\cdot\mathbf{w})\mathbf{v}$$

giving

$$\nabla\cdot(\mathbf{v}\otimes(\mathbf{v}-\boldsymbol{\vartheta})) = ((\mathbf{v}-\boldsymbol{\vartheta})\cdot\nabla)\mathbf{v} + (\nabla\cdot(\mathbf{v}-\boldsymbol{\vartheta}))\mathbf{v}.$$

Using the fact that $\nabla \cdot \mathbf{v} = 0$, we obtain the following equality:

$$[\nabla \cdot (\mathbf{v} \otimes (\mathbf{v} - \boldsymbol{\vartheta}))] \cdot \mathbf{w} = [((\mathbf{v} - \boldsymbol{\vartheta}) \cdot \nabla)\mathbf{v}] \cdot \mathbf{w} - (\nabla \cdot \boldsymbol{\vartheta})(\mathbf{v} \cdot \mathbf{w}).$$

We observe that the second and third terms in [3.14] can be replaced by using the following expression:

$$\rho \int_{\Omega_t} [\nabla \cdot (\mathbf{v} \otimes (\mathbf{v} - \boldsymbol{\vartheta}))] \cdot \mathbf{w} \, d\mathbf{x}.$$

The description "conservative" refers to the fact that the above term can be written as a function of the divergence of an expression.

3.4. Discretization of order one in time for the weak conservative formulation

In this section, we shall use the formalism presented in [NOB 01] and [QUA 04]. Since we are working with non-homogeneous boundary conditions for the velocity, the non-linear term $\int_{\Omega_t} [(\mathbf{u} \cdot \nabla)\mathbf{v}] \cdot \mathbf{w} \, d\mathbf{x}$ does not cancel if the second and third arguments are equal. Hence, we shall replace it.

Consider arbitrary $\mathbf{u}, \mathbf{v}, \mathbf{w} \in (H^1(\Omega_t))^2$. We have

$$[(\mathbf{u} \cdot \nabla)\mathbf{v}] \cdot \mathbf{w} = (u_1 \frac{\partial v_1}{\partial x_1} + u_2 \frac{\partial v_1}{\partial x_2}) w_1 + (u_1 \frac{\partial v_2}{\partial x_1} + u_2 \frac{\partial v_2}{\partial x_2}) w_2$$

and

$$[(\mathbf{u} \cdot \nabla)\mathbf{w}] \cdot \mathbf{v} = (u_1 \frac{\partial w_1}{\partial x_1} + u_2 \frac{\partial w_1}{\partial x_2}) v_1 + (u_1 \frac{\partial w_2}{\partial x_1} + u_2 \frac{\partial w_2}{\partial x_2}) v_2$$

which gives

$$\begin{aligned}
[(\mathbf{u} \cdot \nabla)\mathbf{v}] \cdot \mathbf{w} + [(\mathbf{u} \cdot \nabla)\mathbf{w}] \cdot \mathbf{v} &= u_1 (\frac{\partial v_1}{\partial x_1} w_1 + \frac{\partial v_2}{\partial x_1} w_2 + \frac{\partial w_1}{\partial x_1} v_1 + \frac{\partial w_2}{\partial x_1} v_2) \\
&\quad + u_2 (\frac{\partial v_1}{\partial x_2} w_1 + \frac{\partial v_2}{\partial x_2} w_2 + \frac{\partial w_1}{\partial x_2} v_1 + \frac{\partial w_2}{\partial x_2} v_2) \\
&= \mathbf{u} \cdot \nabla(\mathbf{v} \cdot \mathbf{w}). \qquad [3.17]
\end{aligned}$$

Using Green's formula (see theorem A.5),

$$\int_{\Omega_t} \mathbf{u} \cdot \nabla h \, d\mathbf{x} + \int_{\Omega_t} (\nabla \cdot \mathbf{u}) h \, d\mathbf{x} = \int_{\partial\Omega_t} (\mathbf{u} \cdot \mathbf{n}) h \, ds$$

for $h = \mathbf{v} \cdot \mathbf{w} \in H^1(\Omega_t)$, we obtain the following equality:

$$\int_{\Omega_t} \mathbf{u} \cdot \nabla(\mathbf{v} \cdot \mathbf{w}) \, d\mathbf{x} + \int_{\Omega_t} (\nabla \cdot \mathbf{u})(\mathbf{v} \cdot \mathbf{w}) \, d\mathbf{x} = \int_{\partial\Omega_t} (\mathbf{u} \cdot \mathbf{n})(\mathbf{v} \cdot \mathbf{w}) \, ds$$

and taking account of [3.17], we have

$$\int_{\Omega_t} [(\mathbf{u} \cdot \nabla)\mathbf{v}] \cdot \mathbf{w} \, d\mathbf{x} + \int_{\Omega_t} [(\mathbf{u} \cdot \nabla)\mathbf{w}] \cdot \mathbf{v} \, d\mathbf{x} + \int_{\Omega_t} (\nabla \cdot \mathbf{u})(\mathbf{v} \cdot \mathbf{w}) \, d\mathbf{x}$$
$$= \int_{\partial\Omega_t} (\mathbf{u} \cdot \mathbf{n})(\mathbf{v} \cdot \mathbf{w}) \, ds. \qquad [3.18]$$

We introduce the trilinear form $\tilde{c} : (H^1(\Omega_t))^2 \times (H^1(\Omega_t))^2 \times (H^1(\Omega_t))^2 \to \mathbb{R}$

$$\tilde{c}(\mathbf{u}, \mathbf{v}, \mathbf{w}) \overset{déf}{=} \frac{1}{2} \int_{\Omega_t} [(\mathbf{u} \cdot \nabla)\mathbf{v}] \cdot \mathbf{w} \, d\mathbf{x} - \frac{1}{2} \int_{\Omega_t} [(\mathbf{u} \cdot \nabla)\mathbf{w}] \cdot \mathbf{v} \, d\mathbf{x}$$

which has the property

$$\tilde{c}(\mathbf{u}, \mathbf{w}, \mathbf{w}) = 0, \quad \forall \mathbf{u}, \mathbf{w} \in (H^1(\Omega_t))^2. \qquad [3.19]$$

Taking the second term from [3.18] and replacing it in the definition of $\tilde{c}$, we obtain the following equality:

$$\tilde{c}(\mathbf{u}, \mathbf{v}, \mathbf{w}) = \frac{1}{2} \int_{\Omega_t} [(\mathbf{u} \cdot \nabla)\mathbf{v}] \cdot \mathbf{w} \, d\mathbf{x}$$
$$+ \frac{1}{2} \left(\int_{\Omega_t} [(\mathbf{u} \cdot \nabla)\mathbf{v}] \cdot \mathbf{w} \, d\mathbf{x} + \int_{\Omega_t} (\nabla \cdot \mathbf{u})(\mathbf{v} \cdot \mathbf{w}) \, d\mathbf{x} - \int_{\partial\Omega_t} (\mathbf{u} \cdot \mathbf{n})(\mathbf{v} \cdot \mathbf{w}) \, ds \right)$$
$$= \int_{\Omega_t} [(\mathbf{u} \cdot \nabla)\mathbf{v}] \cdot \mathbf{w} \, d\mathbf{x} + \frac{1}{2} \int_{\Omega_t} (\nabla \cdot \mathbf{u})(\mathbf{v} \cdot \mathbf{w}) \, d\mathbf{x} - \frac{1}{2} \int_{\partial\Omega_t} (\mathbf{u} \cdot \mathbf{n})(\mathbf{v} \cdot \mathbf{w}) \, ds.$$

If

$$\nabla \cdot \mathbf{v} = 0, \quad \text{in } \Omega_t, \tag{3.20}$$

$$\mathbf{w} = 0, \quad \text{on } \Gamma_D \tag{3.21}$$

$$\boldsymbol{\vartheta} = 0, \quad \text{on } \Gamma_N \cup \Gamma_D \tag{3.22}$$

$$\boldsymbol{\vartheta} \cdot \mathbf{n} = \mathbf{g} \cdot \mathbf{n}, \quad \text{on } \Gamma_t \tag{3.23}$$

then

$$\begin{aligned}
\int_{\Omega_t} [((\mathbf{v} - \boldsymbol{\vartheta}) \cdot \nabla)\mathbf{v}] \cdot \mathbf{w}\, d\mathbf{x} &= \tilde{c}(\mathbf{v} - \boldsymbol{\vartheta}, \mathbf{v}, \mathbf{w}) - \frac{1}{2} \int_{\Omega_t} (\nabla \cdot (\mathbf{v} - \boldsymbol{\vartheta}))(\mathbf{v} \cdot \mathbf{w})\, d\mathbf{x} \\
&\quad + \frac{1}{2} \int_{\partial\Omega_t} ((\mathbf{v} - \boldsymbol{\vartheta}) \cdot \mathbf{n})(\mathbf{v} \cdot \mathbf{w})\, ds \\
&= \tilde{c}(\mathbf{v} - \boldsymbol{\vartheta}, \mathbf{v}, \mathbf{w}) + \frac{1}{2} \int_{\Omega_t} (\nabla \cdot \boldsymbol{\vartheta})(\mathbf{v} \cdot \mathbf{w})\, d\mathbf{x} \\
&\quad + \frac{1}{2} \int_{\Gamma_t \cup \Gamma_N \cup \Gamma_D} ((\mathbf{v} - \boldsymbol{\vartheta}) \cdot \mathbf{n})(\mathbf{v} \cdot \mathbf{w})\, ds \\
&= \tilde{c}(\mathbf{v} - \boldsymbol{\vartheta}, \mathbf{v}, \mathbf{w}) + \frac{1}{2} \int_{\Omega_t} (\nabla \cdot \boldsymbol{\vartheta})(\mathbf{v} \cdot \mathbf{w})\, d\mathbf{x} \\
&\quad + \frac{1}{2} \int_{\Gamma_N} (\mathbf{v} \cdot \mathbf{n})(\mathbf{v} \cdot \mathbf{w})\, ds.
\end{aligned}$$

Equation [3.14] is equivalent to

$$\begin{aligned}
&\rho \frac{d}{dt} \int_{\Omega_t} \mathbf{v} \cdot \mathbf{w}\, d\mathbf{x} - \rho \frac{1}{2} \int_{\Omega_t} (\nabla \cdot \boldsymbol{\vartheta})(\mathbf{v} \cdot \mathbf{w})\, d\mathbf{x} + \rho \tilde{c}(\mathbf{v} - \boldsymbol{\vartheta}, \mathbf{v}, \mathbf{w}) \\
&+ \rho \frac{1}{2} \int_{\Gamma_N} (\mathbf{v} \cdot \mathbf{n})(\mathbf{v} \cdot \mathbf{w})\, ds + \int_{\Omega_t} 2\mu\, \boldsymbol{\epsilon}(\mathbf{v}) : \boldsymbol{\epsilon}(\mathbf{w})\, d\mathbf{x} \\
&- \int_{\Omega_t} (\nabla \cdot \mathbf{w})\, p\, d\mathbf{x} = \int_{\Omega_t} \mathbf{f} \cdot \mathbf{w}\, d\mathbf{x} + \int_{\Gamma_N} \mathbf{h} \cdot \mathbf{w}\, ds,
\end{aligned} \tag{3.24}$$

under hypotheses [3.22] and [3.23].

REMARK 3.2.– *We are working with an ALE map that satisfies* $\mathcal{A}_t(\widehat{\mathbf{x}}) = \widehat{\mathbf{x}}$, *for every* $\widehat{\mathbf{x}} \in \Gamma_N \cup \Gamma_D$, *which implies [3.22]. We have* $\mathcal{A}_t(\widehat{\Gamma}) = \Gamma_t$. *If* $\mathcal{A}_t : \widehat{\Gamma} \to \Gamma_t$ *coincides with this Lagrangian description of the boundary* Γ_t, i.e. $\mathcal{A}_t(\mathbf{X}) = \mathbf{X} + \mathbf{U}(\mathbf{X}, t)$, *for every* $\mathbf{X} \in \widehat{\Gamma}$, *then*

$$\boldsymbol{\vartheta}(\mathcal{A}_t(\mathbf{X}), t) = \frac{\partial \mathbf{U}}{\partial t}(\mathbf{X}, t).$$

In fluid-structure interaction problems, we use the following boundary conditions for fluid velocities at the interface Γ_t:

$$\mathbf{g}(\mathbf{X} + \mathbf{U}(\mathbf{X}, t), t) = \frac{\partial \mathbf{U}}{\partial t}(\mathbf{X}, t), \quad \mathbf{X} \in \widehat{\Gamma}$$

thus $\mathbf{g} = \boldsymbol{\vartheta}$ *on* Γ_t, *which implies [3.23]. The ALE maps obtained by harmonic or explicit extensions introduced in a previous section satisfy [3.22] and [3.23].*

We divide the interval $[0, T]$ into $N \in \mathbb{N}^*$ sub-intervals each of length $\Delta t = T/N$ and we note that, $t_n = n\Delta t$, $0 \leq n \leq N$, $\Omega_n = \Omega_{t_n}$, $\Gamma_n = \Gamma_{t_n}$, $\mathcal{A}_n = \mathcal{A}_{t_n}$. We define $\boldsymbol{\vartheta}^n \in (H^1(\Omega_n))^2$ by:

$$\boldsymbol{\vartheta}^n(\mathcal{A}_n(\widehat{\mathbf{x}})) = \frac{\mathcal{A}_n(\widehat{\mathbf{x}}) - \mathcal{A}_{n-1}(\widehat{\mathbf{x}})}{\Delta t}, \quad 1 \leq n \leq N, \text{and } \boldsymbol{\vartheta}^0 = 0.$$

Additionally, for $0 \leq n \leq N-1$ we define $\mathcal{A}^{n,n+1} : \overline{\widehat{\Omega}} \times [t_n, t_{n+1}] \to \mathbb{R}^2$ by:

$$\mathcal{A}^{n,n+1}(\widehat{\mathbf{x}}, t) = \mathcal{A}_t^{n,n+1}(\widehat{\mathbf{x}}) = \mathcal{A}_n(\widehat{\mathbf{x}})\frac{t_{n+1} - t}{\Delta t} + \mathcal{A}_{n+1}(\widehat{\mathbf{x}})\frac{t - t_n}{\Delta t},$$

$t_{n+1/2} = \frac{t_n + t_{n+1}}{2}$ and

$$\mathcal{A}_{n+1/2}(\widehat{\mathbf{x}}) = \mathcal{A}^{n,n+1}(\widehat{\mathbf{x}}, t_{n+1/2}) = \frac{1}{2}\mathcal{A}_n(\widehat{\mathbf{x}}) + \frac{1}{2}\mathcal{A}_{n+1}(\widehat{\mathbf{x}}).$$

We have

$$\widehat{\boldsymbol{\vartheta}}^{n,n+1}(\widehat{\mathbf{x}}, t) = \frac{\partial \mathcal{A}^{n,n+1}}{\partial t}(\widehat{\mathbf{x}}, t) = \frac{\mathcal{A}_{n+1}(\widehat{\mathbf{x}}) - \mathcal{A}_n(\widehat{\mathbf{x}})}{\Delta t}, \quad \forall t \in [t_n, t_{n+1}]. \qquad [3.25]$$

For fixed $\widehat{\mathbf{x}}$, we observe that $\widehat{\boldsymbol{\vartheta}}^{n,n+1}(\widehat{\mathbf{x}}, t)$ is constant for $t \in [t_n, t_{n+1}]$.

Let $\Omega_{n+1/2} = \mathcal{A}_{n+1/2}(\widehat{\Omega})$. We introduce $\boldsymbol{\vartheta}^{n+1/2} \in (H^1(\Omega_{n+1/2}))^2$, which is defined by:

$$\boldsymbol{\vartheta}^{n+1/2}(\mathcal{A}_{n+1/2}(\widehat{\mathbf{x}})) = \frac{\mathcal{A}_{n+1}(\widehat{\mathbf{x}}) - \mathcal{A}_n(\widehat{\mathbf{x}})}{\Delta t}.$$

3.4.1. *A scheme of order one in time for the conservative formulation*

Find $\mathbf{v}^{n+1} \in (H^1(\Omega_{n+1}))^2$ and $p^{n+1} \in L^2(\Omega_{n+1})$, such that $\mathbf{v}^{n+1} = \boldsymbol{\vartheta}^{n+1}$ on Γ_{n+1}, $\mathbf{v}^{n+1} = 0$ on Γ_D and $\forall \mathbf{w} \in W$ and $\forall q \in Q$, we have

$$\begin{aligned}
&\rho\frac{1}{\Delta t}\int_{\Omega_{n+1}} \mathbf{v}^{n+1}\cdot\mathbf{w}\,d\mathbf{x} - \rho\frac{1}{\Delta t}\int_{\Omega_n} \mathbf{v}^{n}\cdot\mathbf{w}\,d\mathbf{x} \\
&-\rho\frac{1}{2}\int_{\Omega_{n+1/2}} (\nabla\cdot\boldsymbol{\vartheta}^{n+1/2})(\mathbf{v}^{n+1}\cdot\mathbf{w})\,d\mathbf{x} + \rho\tilde{c}(\mathbf{v}^n - \boldsymbol{\vartheta}^n, \mathbf{v}^{n+1}, \mathbf{w}) \\
&+\rho\frac{1}{2}\int_{\Gamma_N} (\mathbf{v}^n\cdot\mathbf{n})(\mathbf{v}^{n+1}\cdot\mathbf{w})\,ds + \int_{\Omega_{n+1}} 2\mu\,\boldsymbol{\epsilon}\left(\mathbf{v}^{n+1}\right) : \boldsymbol{\epsilon}(\mathbf{w})\,d\mathbf{x} \\
&-\int_{\Omega_{n+1}} (\nabla\cdot\mathbf{w})\,p^{n+1}\,d\mathbf{x} \\
&= \int_{\Omega_{n+1}} \mathbf{f}^{n+1}\cdot\mathbf{w}\,d\mathbf{x} + \int_{\Gamma_N} \mathbf{h}^{n+1}\cdot\mathbf{w}\,ds, \qquad [3.26]
\end{aligned}$$

$$-\int_{\Omega_{n+1}} \left(\nabla\cdot\mathbf{v}^{n+1}\right) q\,d\mathbf{x} = 0. \qquad [3.27]$$

We initialize with $\mathbf{v}^0 = \mathbf{v}_0$ and $\boldsymbol{\vartheta}^0 = 0$ on Ω_0.

The meaning of some of these integrals should be clarified, for example:

$$\int_{\Omega_{n+1}} \mathbf{v}^{n+1}\cdot\mathbf{w}\,d\mathbf{x} = \int_{\Omega_{n+1}} \mathbf{v}^{n+1}(\mathbf{x})\cdot\mathbf{w}(\mathbf{x}, t_{n+1})\,d\mathbf{x}$$

$$\int_{\Omega_n} \mathbf{v}^{n}\cdot\mathbf{w}\,d\mathbf{x} = \int_{\Omega_n} \mathbf{v}^{n}(\mathbf{x})\cdot\mathbf{w}(\mathbf{x}, t_n)\,d\mathbf{x}$$

$$\int_{\Omega_{n+1/2}} (\nabla \cdot \boldsymbol{\vartheta}^{n+1/2})(\mathbf{v}^{n+1} \cdot \mathbf{w})\, d\mathbf{x}$$
$$= \int_{\Omega_{n+1/2}} (\nabla \cdot \boldsymbol{\vartheta}^{n+1/2}) \left(\mathbf{v}^{n+1} \circ \mathcal{A}_{n+1} \circ \mathcal{A}_{n+1/2}^{-1}\right) \cdot \mathbf{w}(\cdot, t_{n+1/2})\, d\mathbf{x}$$

and similarly for $\tilde{c}(\mathbf{v}^n - \boldsymbol{\vartheta}^n, \mathbf{v}^{n+1}, \mathbf{w})$ where the domain of integration is Ω_{n+1}, but $\mathbf{v}^n$ and $\boldsymbol{\vartheta}^n$ are defined in Ω_n.

3.4.2. *Geometric conservation law*

PROPOSITION 3.6.– *Let* $\widehat{\psi} \in C^0(\widehat{\Omega})$. *Hence*

$$\int_{\Omega_{n+1}} \widehat{\psi} \circ \mathcal{A}_{n+1}^{-1}\, d\mathbf{x} - \int_{\Omega_n} \widehat{\psi} \circ \mathcal{A}_n^{-1}\, d\mathbf{x}$$
$$= \Delta t \int_{\Omega_{n+1/2}} (\nabla \cdot \boldsymbol{\vartheta}^{n+1/2}) \widehat{\psi} \circ \mathcal{A}_{n+1/2}^{-1}\, d\mathbf{x}. \qquad [3.28]$$

DEMONSTRATION 3.4.– For $\widehat{\mathbf{x}} \in \widehat{\Omega}$ and $t \in [t_n, t_{n+1}]$, let

$$\widehat{J}^{n,n+1}(\widehat{\mathbf{x}}, t) = \det\left(\nabla_{\widehat{\mathbf{x}}} \mathcal{A}^{n,n+1}(\mathbf{x}, t)\right)$$
$$= \det\left(\nabla_{\widehat{\mathbf{x}}} \mathcal{A}_n(\widehat{\mathbf{x}}) \frac{t_{n+1} - t}{\Delta t} + \nabla_{\widehat{\mathbf{x}}} \mathcal{A}_{n+1}(\widehat{\mathbf{x}}) \frac{t - t_n}{\Delta t}\right).$$

Since we are in dimension two, the above determinant will be a polynomial of degree two in t

$$\widehat{J}^{n,n+1}(\widehat{\mathbf{x}}, t) = a_2(\widehat{\mathbf{x}})t^2 + a_1(\widehat{\mathbf{x}})t + a_0(\widehat{\mathbf{x}}).$$

We have the following property, specific to polynomials of degree two in t

$$\widehat{J}^{n,n+1}(\widehat{\mathbf{x}}, t_{n+1}) - \widehat{J}^{n,n+1}(\widehat{\mathbf{x}}, t_n) = \Delta t \frac{\partial \widehat{J}^{n,n+1}}{\partial t}(\widehat{\mathbf{x}}, t_{n+1/2}). \qquad [3.29]$$

Using the ALE dilatation formula, proposition 3.2, we obtain the following equality:

$$\frac{\partial \widehat{J}^{n,n+1}}{\partial t}(\widehat{\mathbf{x}}, t_{n+1/2}) = \widehat{J}^{n,n+1}(\widehat{\mathbf{x}}, t_{n+1/2})(\nabla \cdot \boldsymbol{\vartheta}^{n+1/2})(\mathcal{A}_{n+1/2}(\widehat{\mathbf{x}})).$$

Thus,

$$\int_{\widehat{\Omega}} \widehat{\psi}(\widehat{\mathbf{x}}) \widehat{J}^{n,n+1}(\widehat{\mathbf{x}}, t_{n+1})\, d\widehat{\mathbf{x}} - \int_{\widehat{\Omega}} \widehat{\psi}(\widehat{\mathbf{x}}) \widehat{J}^{n,n+1}(\widehat{\mathbf{x}}, t_n)\, d\widehat{\mathbf{x}}$$

$$= \Delta t \int_{\widehat{\Omega}} \widehat{\psi}(\widehat{\mathbf{x}})(\nabla \cdot \boldsymbol{\vartheta}^{n+1/2})(\mathcal{A}_{n+1/2}(\widehat{\mathbf{x}}))\widehat{J}^{n,n+1}(\widehat{\mathbf{x}}, t_{n+1/2})\, d\widehat{\mathbf{x}}$$

which is equivalent to [3.28] after the changes in variables. □

REMARK 3.3.– *Relation [3.28] is a* geometric conservation law, *valid independently of* $\widehat{\psi}$. *It is a discrete version of the equality*

$$\frac{d}{dt} \int_{\Omega_t} \widehat{\psi} \circ \mathcal{A}_t^{-1}(\mathbf{x})\, d\mathbf{x} = \int_{\Omega_t} \widehat{\psi} \circ \mathcal{A}_t^{-1}(\mathbf{x})(\nabla \cdot \boldsymbol{\vartheta})(\mathbf{x}, t)\, d\mathbf{x}$$

which we can demonstrate using ALE dilatation formulation. Equality [3.29] can be seen as the mid-point quadrature formula, which is exact for polynomials of degree one:

$$\int_{t_n}^{t_{n+1}} f(t)dt = \Delta t f(t_{n+1/2})$$

for $f(t) = \frac{\partial \widehat{J}^{n,n+1}}{\partial t}(\widehat{\mathbf{x}}, t)$. *If we work in dimension three,* $\widehat{J}^{n,n+1}(\widehat{\mathbf{x}}, t)$ *will be a polynomial of degree three in t, so a quadrature formula that is exact for polynomials of at least degree two is required.*

We note $|\mathbf{v}^{n+1}|^2 = \mathbf{v}^{n+1} \cdot \mathbf{v}^{n+1}$ and

$$\begin{aligned} E^{n+1} = & \frac{1}{\Delta t} \int_{\Omega_{n+1}} \mathbf{v}^{n+1} \cdot \mathbf{v}^{n+1}\, d\mathbf{x} - \frac{1}{\Delta t} \int_{\Omega_n} \mathbf{v}^n \cdot \mathbf{v}^{n+1}\, d\mathbf{x} \\ & - \frac{1}{2} \int_{\Omega_{n+1/2}} (\nabla \cdot \boldsymbol{\vartheta}^{n+1/2})(\mathbf{v}^{n+1} \cdot \mathbf{v}^{n+1})\, d\mathbf{x} \\ & + \tilde{c}(\mathbf{v}^n - \boldsymbol{\vartheta}^n, \mathbf{v}^{n+1}, \mathbf{v}^{n+1}) + \frac{1}{2} \int_{\Gamma_N} (\mathbf{v}^n \cdot \mathbf{n})(\mathbf{v}^{n+1} \cdot \mathbf{v}^{n+1})\, ds. \end{aligned}$$

PROPOSITION 3.7.– *If* $\mathbf{v}^{n+1} \in (H^1(\Omega_{n+1}))^2$, *then*

$$\frac{1}{2\Delta t} \int_{\Omega_{n+1}} |\mathbf{v}^{n+1}|^2 d\mathbf{x} - \frac{1}{2\Delta t} \int_{\Omega_n} |\mathbf{v}^n|^2 d\mathbf{x} + \frac{1}{2} \int_{\Gamma_N} (\mathbf{v}^n \cdot \mathbf{n}) |\mathbf{v}^{n+1}|^2 ds$$

$$\leq E^{n+1}. \qquad [3.30]$$

DEMONSTRATION 3.5.– In light of property [3.19], we have $\tilde{c}(\mathbf{v}^n - \boldsymbol{\vartheta}^n, \mathbf{v}^{n+1}, \mathbf{v}^{n+1}) = 0$. Taking account of the inequality

$$\int_{\Omega_n} \mathbf{v}^n \cdot \mathbf{v}^{n+1}\, d\mathbf{x} \leq \frac{1}{2}\int_{\Omega_n} |\mathbf{v}^{n+1}|^2 d\mathbf{x} + \frac{1}{2}\int_{\Omega_n} |\mathbf{v}^n|^2 d\mathbf{x}$$

we obtain the following inequality:

$$E^{n+1} \geq \frac{1}{\Delta t}\int_{\Omega_{n+1}} |\mathbf{v}^{n+1}|^2 d\mathbf{x} - \frac{1}{2\Delta t}\int_{\Omega_n} |\mathbf{v}^{n+1}|^2 d\mathbf{x} - \frac{1}{2\Delta t}\int_{\Omega_n} |\mathbf{v}^n|^2 d\mathbf{x}$$
$$-\frac{1}{2}\int_{\Omega_{n+1/2}} (\nabla \cdot \boldsymbol{\vartheta}^{n+1/2})|\mathbf{v}^{n+1}|^2\, d\mathbf{x} + \frac{1}{2}\int_{\Gamma_N} (\mathbf{v}^n \cdot \mathbf{n})|\mathbf{v}^{n+1}|^2 d\mathbf{s}.$$

We now use [3.28] with $\psi = |\mathbf{v}^{n+1}|^2$ and we have

$$\frac{1}{2\Delta t}\int_{\Omega_{n+1}} |\mathbf{v}^{n+1}|^2 d\mathbf{x} - \frac{1}{2\Delta t}\int_{\Omega_n} |\mathbf{v}^{n+1}|^2 d\mathbf{x} = \frac{1}{2}\int_{\Omega_{n+1/2}} (\nabla \cdot \boldsymbol{\vartheta}^{n+1/2})|\mathbf{v}^{n+1}|^2\, d\mathbf{x}.$$

giving [3.30]. □

3.5. Stabilized discretization of order one in time for the non-conservative formulation

We assume that the following are known at time t_n:

– the domain Ω_n, an open, connected, and bounded set and its boundary $\partial\Omega_n = \overline{\Gamma}_D \cup \overline{\Gamma}_N \cup \overline{\Gamma}_n$ is Lipschitz,

– the fluid velocity $\mathbf{v}^n \in (H^1(\Omega_n))^2$, such that $\nabla \cdot \mathbf{v}^n = 0$ in Ω_n, $\mathbf{v}^n = \mathbf{g}^n$ on Γ_n, $\mathbf{v}^n = 0$ on Γ_D,

– the velocity of the domain $\boldsymbol{\vartheta}^n \in \left(H^1(\Omega_n) \cap C^0(\overline{\Omega}_n)\right)^2$, such that $\boldsymbol{\vartheta}^n = \mathbf{g}^n$ on Γ_n, $\boldsymbol{\vartheta}^n = 0$ on $\Gamma_D \cup \Gamma_N$.

We shall choose $\widehat{\Omega} = \Omega_n$ and we define $\mathcal{A}_{n+1} : \Omega_n \rightarrow \mathbb{R}^2$ by:

$$\mathcal{A}_{n+1}(\widehat{\mathbf{x}}) = \widehat{\mathbf{x}} + \Delta t \boldsymbol{\vartheta}^n(\widehat{\mathbf{x}}).$$

We note $\Omega_{n+1} = \mathcal{A}_{n+1}(\Omega_n)$, $\Gamma_{n+1} = \mathcal{A}_{n+1}(\Gamma_n)$ and we have $\mathcal{A}_{n+1}(\widehat{\mathbf{x}}) = \widehat{\mathbf{x}}$ on $\Gamma_D \cup \Gamma_N$.

The Jacobian is

$$\widehat{J}_{n+1}(\widehat{\mathbf{x}}) = \det(\nabla_{\widehat{\mathbf{x}}}\mathcal{A}_{n+1}(\widehat{\mathbf{x}})) = 1 + \Delta t \nabla_{\widehat{\mathbf{x}}} \cdot \boldsymbol{\vartheta}^n(\widehat{\mathbf{x}}) + (\Delta t)^2 \det(\nabla_{\widehat{\mathbf{x}}}\boldsymbol{\vartheta}^n(\widehat{\mathbf{x}})). \quad [3.31]$$

We are looking to find, at time t_{n+1} the fluid velocity $\mathbf{v}^{n+1} \in (H^1(\Omega_{n+1}))^2$, such that $\mathbf{v}^{n+1} = \mathbf{g}^{n+1}$ on Γ_{n+1}, $\mathbf{v}^{n+1} = 0$ on Γ_D and the fluid pressure $p^{n+1} \in L^2(\Omega_{n+1})$ satisfying

$$\begin{aligned}
&\rho \int_{\Omega_n} \frac{\widehat{\mathbf{v}}^{n+1} - \mathbf{v}^n}{\Delta t} \cdot \widehat{\mathbf{w}}\, d\widehat{\mathbf{x}} + \rho \int_{\Omega_n} [((\mathbf{v}^n - \boldsymbol{\vartheta}^n) \cdot \nabla_{\widehat{\mathbf{x}}})\widehat{\mathbf{v}}^{n+1}] \cdot \widehat{\mathbf{w}}\, d\widehat{\mathbf{x}} \\
&+\rho \frac{\Delta t}{2} \int_{\Omega_n} \det(\nabla_{\widehat{\mathbf{x}}}\boldsymbol{\vartheta}^n)\widehat{\mathbf{v}}^{n+1} \cdot \widehat{\mathbf{w}}\, d\widehat{\mathbf{x}} + \int_{\Omega_{n+1}} 2\mu\, \boldsymbol{\epsilon}\left(\mathbf{v}^{n+1}\right) : \boldsymbol{\epsilon}(\mathbf{w})\, d\mathbf{x} \\
&- \int_{\Omega_{n+1}} (\nabla \cdot \mathbf{w})\, p^{n+1}\, d\mathbf{x} \\
&= \int_{\Omega_{n+1}} \mathbf{f}^{n+1} \cdot \mathbf{w}\, d\mathbf{x} + \int_{\Gamma_N} \mathbf{h}^{n+1} \cdot \mathbf{w}\, ds, \quad [3.32] \\
&- \int_{\Omega_{n+1}} \left(\nabla \cdot \mathbf{v}^{n+1}\right) q\, d\mathbf{x} = 0 \quad [3.33]
\end{aligned}$$

for every $\widehat{\mathbf{w}}$ in $(H^1(\Omega_n))^2$, such that $\widehat{\mathbf{w}} = 0$ on $\Gamma_n \cup \Gamma_D$, and every q in $L^2(\Omega_{n+1})$. We have used the convention $\widehat{\mathbf{w}} = \mathbf{w} \circ \mathcal{A}_{n+1}$.

We can construct $\boldsymbol{\vartheta}^{n+1} : \overline{\Omega}_{n+1} \to \mathbb{R}^2$ by harmonic extension

$$\Delta\boldsymbol{\vartheta}^{n+1} = 0 \text{ in } \Omega_{n+1},\ \boldsymbol{\vartheta}^{n+1} = 0 \text{ on } \Gamma_N \cup \Gamma_D,\ \boldsymbol{\vartheta}^{n+1} = \mathbf{g}^{n+1} \text{ on } \Gamma_{n+1}. \quad [3.34]$$

We initialize with $\mathbf{v}^0 = \mathbf{v}_0$ and $\Delta\boldsymbol{\vartheta}^0 = 0$ on Ω_0, $\boldsymbol{\vartheta}^0 = 0$ on $\Gamma_N \cup \Gamma_D$, $\boldsymbol{\vartheta}^0 = \mathbf{g}^0$ on Γ_0.

We note $|\mathbf{v}^{n+1}|^2 = \mathbf{v}^{n+1} \cdot \mathbf{v}^{n+1}$ and

$$\begin{aligned}
E^{n+1} = &\int_{\Omega_n} \frac{\widehat{\mathbf{v}}^{n+1} - \mathbf{v}^n}{\Delta t} \cdot \widehat{\mathbf{v}}^{n+1}\, d\widehat{\mathbf{x}} + \int_{\Omega_n} [((\mathbf{v}^n - \boldsymbol{\vartheta}^n) \cdot \nabla_{\widehat{\mathbf{x}}})\widehat{\mathbf{v}}^{n+1}] \cdot \widehat{\mathbf{v}}^{n+1}\, d\widehat{\mathbf{x}} \\
&+ \frac{\Delta t}{2} \int_{\Omega_n} \det(\nabla_{\widehat{\mathbf{x}}}\boldsymbol{\vartheta}^n)\widehat{\mathbf{v}}^{n+1} \cdot \widehat{\mathbf{v}}^{n+1}\, d\widehat{\mathbf{x}}.
\end{aligned}$$

PROPOSITION 3.8.– *If:* $\mathbf{v}^{n+1} \in (H^1(\Omega_{n+1}))^2$, $\mathbf{v}^{n+1} = 0$ *on* Γ_D, $\mathbf{v}^n \in (H^1(\Omega_n))^2$, $\nabla_{\widehat{\mathbf{x}}} \cdot \mathbf{v}^n = 0$ *in* Ω_n, $\mathbf{v}^n = \mathbf{g}^n$ *on* Γ_n, $\mathbf{v}^n = 0$ *on* Γ_D, $\boldsymbol{\vartheta}^n \in \left(H^1(\Omega_n) \cap C^0(\overline{\Omega}_n)\right)^2$, $\boldsymbol{\vartheta}^n = \mathbf{g}^n$ *on* Γ_n, $\boldsymbol{\vartheta}^n = 0$ *on* $\Gamma_D \cup \Gamma_N$, *then*

$$\frac{1}{2\Delta t}\int_{\Omega_{n+1}} |\mathbf{v}^{n+1}|^2 d\mathbf{x} - \frac{1}{2\Delta t}\int_{\Omega_n} |\mathbf{v}^n|^2 d\mathbf{x} + \frac{1}{2}\int_{\Gamma_N} (\mathbf{v}^n \cdot \mathbf{n})|\mathbf{v}^{n+1}|^2 ds \\ \leq E^{n+1}. \qquad [3.35]$$

DEMONSTRATION 3.6.– We have

$$\int_{\Omega_n} \mathbf{v}^n \cdot \widehat{\mathbf{v}}^{n+1}\, d\mathbf{x} \leq \frac{1}{2}\int_{\Omega_n} |\mathbf{v}^n|^2 d\mathbf{x} + \frac{1}{2}\int_{\Omega_n} |\widehat{\mathbf{v}}^{n+1}|^2 d\mathbf{x}$$

and by using $[(\mathbf{w} \cdot \nabla)\mathbf{v}] \cdot \mathbf{v} = \frac{1}{2}\mathbf{w} \cdot (\nabla |\mathbf{v}|^2)$, we obtain the following equalities:

$$\int_{\Omega_n} [((\mathbf{v}^n - \boldsymbol{\vartheta}^n) \cdot \nabla_{\widehat{\mathbf{x}}})\widehat{\mathbf{v}}^{n+1}] \cdot \widehat{\mathbf{v}}^{n+1}\, d\widehat{\mathbf{x}} = \frac{1}{2}\int_{\Omega_n} (\mathbf{v}^n - \boldsymbol{\vartheta}^n) \cdot (\nabla_{\widehat{\mathbf{x}}} |\widehat{\mathbf{v}}^{n+1}|^2)\, d\widehat{\mathbf{x}}$$

$$= \frac{1}{2}\int_{\partial\Omega_n} (\mathbf{v}^n - \boldsymbol{\vartheta}^n) \cdot \mathbf{n} |\widehat{\mathbf{v}}^{n+1}|^2 ds - \frac{1}{2}\int_{\Omega_n} \nabla_{\widehat{\mathbf{x}}} \cdot (\mathbf{v}^n - \boldsymbol{\vartheta}^n) |\widehat{\mathbf{v}}^{n+1}|^2 d\widehat{\mathbf{x}}$$

$$= \frac{1}{2}\int_{\Gamma_N} \mathbf{v}^n \cdot \mathbf{n} |\widehat{\mathbf{v}}^{n+1}|^2 ds + \frac{1}{2}\int_{\Omega_n} (\nabla_{\widehat{\mathbf{x}}} \cdot \boldsymbol{\vartheta}^n) |\widehat{\mathbf{v}}^{n+1}|^2 d\widehat{\mathbf{x}}.$$

Hence,

$$E^{n+1} \geq \frac{1}{2\Delta t}\int_{\Omega_n} |\widehat{\mathbf{v}}^{n+1}|^2 d\mathbf{x} - \frac{1}{2\Delta t}\int_{\Omega_n} |\mathbf{v}^n|^2 d\mathbf{x} + \frac{1}{2}\int_{\Gamma_N} \mathbf{v}^n \cdot \mathbf{n} |\mathbf{v}^{n+1}|^2 ds \\ + \frac{1}{2}\int_{\Omega_n} (\nabla_{\widehat{\mathbf{x}}} \cdot \boldsymbol{\vartheta}^n) |\widehat{\mathbf{v}}^{n+1}|^2 d\widehat{\mathbf{x}} + \frac{\Delta t}{2}\int_{\Omega_n} \det(\nabla_{\widehat{\mathbf{x}}} \boldsymbol{\vartheta}^n) |\widehat{\mathbf{v}}^{n+1}|^2\, d\widehat{\mathbf{x}} \\ = \frac{1}{2\Delta t}\int_{\Omega_n} \left(1 + \Delta t(\nabla_{\widehat{\mathbf{x}}} \cdot \boldsymbol{\vartheta}^n) + (\Delta t)^2 \det(\nabla_{\widehat{\mathbf{x}}} \boldsymbol{\vartheta}^n)\right) |\widehat{\mathbf{v}}^{n+1}|^2 d\mathbf{x} \\ - \frac{1}{2\Delta t}\int_{\Omega_n} |\mathbf{v}^n|^2 d\mathbf{x} + \frac{1}{2}\int_{\Gamma_N} \mathbf{v}^n \cdot \mathbf{n} |\mathbf{v}^{n+1}|^2 ds.$$

We now use [3.31] and we obtain [3.35]. □

REMARK 3.4.– *By adding the stabilization term containing* $(\Delta t)\det(\nabla_{\widehat{\mathbf{x}}}\boldsymbol{\vartheta}^n)$ *to [3.32], we do not change the consistency of the scheme. In dimension three, the stabilization term will be*

$$\frac{1}{\Delta t}\left(\widehat{J}_{n+1} - 1 - \Delta t(\nabla_{\widehat{\mathbf{x}}}\cdot\boldsymbol{\vartheta}^n)\right).$$

We need neither a geometric conservation law, nor to evaluate at time $t_{n+1/2}$.

3.6. Finite element discretization in space

For discretization in time, we used the notation: $t_n = n\Delta t$, $\Omega_n = \Omega_{t_n}$, $\mathcal{A}_n = \mathcal{A}_{t_n}$.

We assume that we have a triangulation $\widehat{\mathcal{T}}_h$ of the ALE reference domain $\widehat{\Omega}$.

We shall return to the example presented in Figure 3.1 with $\widehat{\Omega} =]0, L[\times]0, H[$ and the explicit ALE map $\mathcal{A}_t : \overline{\widehat{\Omega}} \to \overline{\Omega}_t$ defined by:

$$\mathcal{A}_t(\widehat{x}_1, \widehat{x}_2) = \left(\widehat{x}_1, \frac{H + \widehat{u}_t(\widehat{x}_1)}{H}\widehat{x}_2\right),$$

with an upper boundary displacement $\widehat{u}_t : [0, T] \to \mathbb{R}$ defined by:

$$\widehat{u}_t(\widehat{x}_1) = \widehat{x}_1^2(L - \widehat{x}_1^2)\alpha(t)$$

where $\alpha : [0, T] \to \mathbb{R}$.

Let $\widehat{K}$ be the triangle with vertices $\widehat{A} = (0, 0)$, $\widehat{B} = (H, H)$, $\widehat{C} = (0, H)$. We shall study the image of $\widehat{K}$ under the map $\mathcal{A}_n$, i.e.

$$K = \mathcal{A}_n(\widehat{K}).$$

The image of the line segment $[\widehat{C}, \widehat{B}] = \{(\widehat{x}_1, H);\ \widehat{x}_1 \in [0, H]\}$ will be the curve

$$\mathcal{A}_n([\widehat{C}, \widehat{B}]) = \{(\widehat{x}_1, H + \widehat{x}_1^2(L - \widehat{x}_1^2)\alpha(t_n));\ \widehat{x}_1 \in [0, H]\},$$

which is not a line segment, and so K is not a triangle! Thus $\mathcal{A}_n(\widehat{\mathcal{T}}_h)$ is not necessarily a triangulation of Ω_n.

In the following, we shall construct a triangulation $\mathcal{T}_h^n$ of Ω_n using $\widehat{\mathcal{T}}_h$ and $\mathcal{A}_n$.

The point A is a vertex of $\mathcal{T}_h^n$ if and only if $\widehat{A} = \mathcal{A}_n^{-1}(A)$ is a vertex of $\widehat{\mathcal{T}}_h$. The triangle $K = \Delta\, ABC \in \mathcal{T}_h^n$ if and only if the triangle $\widehat{K} = \Delta\, \widehat{ABC} \in \widehat{\mathcal{T}}_h$, where $\widehat{A} = \mathcal{A}_n^{-1}(A)$, $\widehat{B} = \mathcal{A}_n^{-1}(B)$, $\widehat{C} = \mathcal{A}_n^{-1}(C)$. In other words, we have moved the vertices, but kept the same connections.

For every $\widehat{K} = \Delta\, \widehat{ABC} \in \widehat{\mathcal{T}}_h$, there exists a unique bijection $\mathcal{A}_{n,h}|_{\widehat{K}} : \widehat{K} \to K$, where $K = \Delta\, ABC$ such that

$$\mathcal{A}_{n,h}|_{\widehat{K}} \in \left(\mathbb{P}_1(\widehat{K})\right)^2.$$

Overall, $\mathcal{A}_{n,h}$ is a continuous map on $\widehat{\mathcal{T}}_h$, with the restriction on each triangle being a polynomial of degree one. We have that $\mathcal{T}_h^n = \mathcal{A}_{n,h}(\widehat{\mathcal{T}}_h)$ is a triangulation of Ω_n.

We have another property for the finite element approximation. For $\ell \in \mathbb{N}$, let

$$\widehat{Q} = \{\widehat{q} \in C^0(\widehat{\mathcal{T}}_h);\ \forall \widehat{K} \in \widehat{\mathcal{T}}_h,\ \widehat{q}|_{\widehat{K}} \in \mathbb{P}_\ell(\widehat{K})\}$$

and

$$Q = \{q : \mathcal{T}_h^n \to \mathbb{R};\ \exists \widehat{q} \in \widehat{Q},\ q = \widehat{q} \circ \mathcal{A}_{n,h}^{-1}\},$$

hence

$$Q = \{q \in C^0(\mathcal{T}_h^n);\ \forall K \in \mathcal{T}_h^n,\ q|_K \in \mathbb{P}_\ell(K)\}.$$

3.7. Numerical tests

We have adapted the benchmark proposed in [NOB 01, pp. 92–94]. The ALE reference domain is $\widehat{\Omega} =]0, H[\times]0, L[$, where $H = 1$ and $L = 6$. We shall note the boundaries of the reference domain with $\widehat{\Gamma}_N =]0, H[\times\{L\}$ and

$\widehat{\Gamma}_D = \partial\widehat{\Omega} \setminus \overline{\widehat{\Gamma}}_N$. Let $\widehat{\mathbf{u}} : \partial\widehat{\Omega} \times [0,T] \to \mathbb{R}^2$ be the displacement of the boundary given by the function

$$\widehat{\mathbf{u}}(\widehat{x}_1, \widehat{x}_2, t) = \left(0,\ 0.4\sin\left(\frac{2\pi t}{10}\right)(\widehat{x}_2 - 0.5)\right)$$

and we denote the domain bounded by the boundary $\mathcal{A}_t(\partial\widehat{\Omega})$ with Ω_t where

$$\begin{aligned}\mathcal{A}_t(\widehat{\mathbf{x}}) &= \widehat{\mathbf{x}} + \widehat{\mathbf{u}}(\widehat{\mathbf{x}}, t)\\ &= \left(\widehat{x}_1,\ \left(1 + 0.4\sin\left(\frac{2\pi t}{10}\right)\right)(\widehat{x}_2 - 0.5) + 0.5\right).\end{aligned}$$

We are looking for the velocity $\mathbf{v}(\cdot,t) = (v_1(\cdot,t), v_2(\cdot,t)) : \overline{\Omega}_t \to \mathbb{R}^2$ and the pressure $p(\cdot,t) : \overline{\Omega}_t \to \mathbb{R}$, such that

$$\rho\left(\frac{\partial \mathbf{v}}{\partial t} + (\mathbf{v}\cdot\nabla)\mathbf{v}\right) - 2\mu\nabla\cdot\epsilon(\mathbf{v}) + \nabla p = \mathbf{f}, \forall t \in]0,T[\ \forall \mathbf{x} \in \Omega_t \qquad [3.36]$$

$$\nabla\cdot\mathbf{v} = 0,\ \forall t \in]0,T[, \forall \mathbf{x} \in \Omega_t \qquad [3.37]$$

$$\mathbf{v} = \mathbf{g},\ \forall t \in]0,T[, \forall \mathbf{x} \in \mathcal{A}_t(\widehat{\Gamma}_D) \qquad [3.38]$$

$$\sigma\mathbf{n} = \mathbf{h},\ \forall t \in]0,T[, \forall \mathbf{x} \in \mathcal{A}_t(\widehat{\Gamma}_N) \qquad [3.39]$$

$$\mathbf{v}(\mathbf{x},0) = \mathbf{v}_0(\mathbf{x}),\ \forall \mathbf{x} \in \Omega_0. \qquad [3.40]$$

We impose the Dirichlet conditions for the velocity on the top, bottom and left-hand boundaries and we impose traction forces on the right-hand boundary.

The exact solution is given as follows:

$$\mathbf{v}_{ext}(x_1, x_2, t) = \left(-\frac{2V(x_1 - L)}{(1+2Vt)}, \frac{2V(x_2 - 0.5)}{(1+2Vt)}\right),$$

$$p_{ext}(x_1, x_2, t) = -\left(\frac{2V(x_1 - L)}{(1+2Vt)}\right)^2$$

where $V = 0.2$. We have used the physical parameters: $\rho = 1, \mu = 1$,

$$\mathbf{f} = (f_1, f_2) = (0, 0), \quad \mathbf{g} = \mathbf{v}_{ext}, \quad \mathbf{h} = (h_1, h_2) = \left(-\frac{4V}{(1+2Vt)}, 0\right)$$

and the initial velocity is $\mathbf{v}_0(x_1, x_2) = \mathbf{v}_{ext}(x_1, x_2, 0)$.

We divide the interval $[0, T]$ into $N \in \mathbb{N}^*$ sub-intervals each of length $\Delta t = T/N$ and we note that $t_n = n\Delta t$, $0 \le n \le N$, $\Omega_n = \Omega_{t_n}$, $\widehat{\mathbf{u}}^n = \widehat{\mathbf{u}}(\cdot, t_n)$.

We shall construct by harmonic extension, $\mathcal{A}_{n+1} : \overline{\widehat{\Omega}} \to \overline{\Omega}_{n+1}$, the discrete ALE map

$$\mathcal{A}_{n+1}(\widehat{\mathbf{x}}) = \widehat{\mathbf{x}} + \widehat{\mathbf{d}}^{n+1}(\widehat{\mathbf{x}})$$

where

$$\Delta \widehat{\mathbf{d}}^{n+1} = 0 \text{ in } \widehat{\Omega}, \quad \widehat{\mathbf{d}}^{n+1} = \widehat{\mathbf{u}}^{n+1} \text{ on } \partial\widehat{\Omega}.$$

We shall define the discrete velocity of the domain at time t_{n+1} by:

$$\widehat{\boldsymbol{\vartheta}}^{n+1} = \frac{\widehat{\mathbf{d}}^{n+1} - \widehat{\mathbf{d}}^n}{\Delta t}, \quad \boldsymbol{\vartheta}^{n+1} = \widehat{\boldsymbol{\vartheta}}^{n+1} \circ \mathcal{A}_{n+1}^{-1}.$$

We shall use a non-stabilized scheme for the non-conservative form: knowing the velocity $\mathbf{v}^n : \Omega_n \to \mathbb{R}^2$, the current domain Ω_{n+1} and the mesh velocity $\boldsymbol{\vartheta}^{n+1} : \Omega_{n+1} \to \mathbb{R}^2$, find the velocity $\mathbf{v}^{n+1} : \Omega_{n+1} \to \mathbb{R}^2$ satisfying $\mathbf{v}^{n+1} = \mathbf{g}^{n+1}$ on $\mathcal{A}_{n+1}(\widehat{\Gamma}_D)$ and the pressure $p^{n+1} : \Omega_{n+1} \to \mathbb{R}$, such that

$$\int_{\Omega_{n+1}} \rho\left(\frac{\mathbf{v}^{n+1} - \mathbf{V}^n}{\Delta t}\right) \cdot \mathbf{w}\, d\mathbf{x} + \int_{\Omega_{n+1}} \rho\left(\left(\left(\mathbf{V}^n - \boldsymbol{\vartheta}^{n+1}\right) \cdot \nabla\right) \mathbf{v}^{n+1}\right) \cdot \mathbf{w}\, d\mathbf{x}$$

$$+ \int_{\Omega_{n+1}} 2\mu\, \boldsymbol{\epsilon}\left(\mathbf{v}^{n+1}\right) : \boldsymbol{\epsilon}(\mathbf{w})\, d\mathbf{x} - \int_{\Omega_{n+1}} (\nabla \cdot \mathbf{w})\, p^{n+1}\, d\mathbf{x}$$

$$= \int_{\Omega_{n+1}} \mathbf{f}^{n+1} \cdot \mathbf{w}\, d\mathbf{x} + \int_{\mathcal{A}_{n+1}(\widehat{\Gamma}_N)} \mathbf{h}^{n+1} \cdot \mathbf{w}\, ds,$$

$$\forall \mathbf{w} \in (H^1(\Omega_{n+1}))^2, \ \mathbf{w} = 0 \text{ sur } \mathcal{A}_{n+1}(\widehat{\Gamma}_D)$$

$$-\int_{\Omega_{n+1}} \left(\nabla \cdot \mathbf{v}^{n+1}\right) q\, d\mathbf{x} = 0, \quad \forall q \in L^2(\Omega_{n+1}),$$

where $\mathbf{V}^n(\mathbf{x}) = \mathbf{v}^n\left(\mathcal{A}_n \circ \mathcal{A}_{n+1}^{-1}(\mathbf{x})\right)$. In contrast to the stabilized scheme for the non-conservative form studied in a previous section, here all of the terms are evaluated on Ω_{n+1}, which facilitates implementation on a computer.

We have conducted the tests in $[0, T = 7]$, with steps in time of $\Delta t = 1/2$, 1/4, 1/8, 1/16, 1/32, as in [NOB 01]. We have used a mesh of 1248 triangles and 695 vertices, and the finite elements $\mathbb{P}_1 + b$ for the velocity and $\mathbb{P}_1$ for the pressure.

The norm

$$||p||_{L^2(0,T;\,L^2(\Omega_t))} = \sqrt{\int_0^T ||p(t)||^2_{0,\Omega_t}}$$

has been approximated by

$$\sqrt{\sum_{n=1}^{N} \Delta t\, ||p^n||^2_{L^2(\Omega_n)}}.$$

Figures 3.2 and 3.3 show good agreement between the calculated and exact values, for velocity and pressure. In applications such as the fluid-structure interaction, the fluid forces acting on the sides of the structure, given by $-\boldsymbol{\sigma}\mathbf{n}$, where $\boldsymbol{\sigma} = -p\mathbf{I} + 2\mu\boldsymbol{\epsilon}(\mathbf{v})$, must be known. Figure 3.4 shows that the differences between the calculated and exact forces acting on the right-hand side boundary are slight.

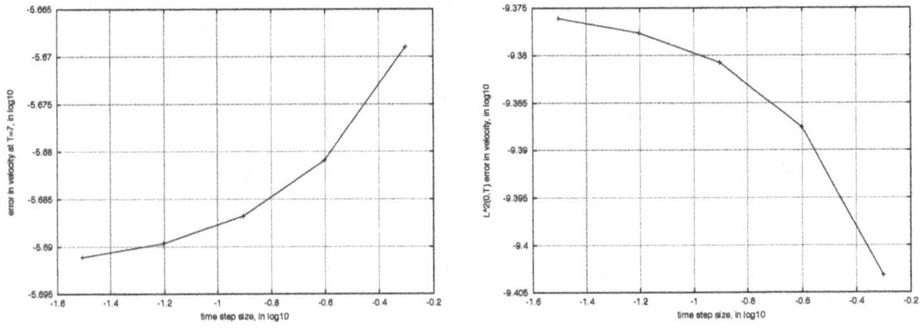

Figure 3.2. *Velocity. The error in the norm $L^2(\Omega_T)$ between the exact and calculated solutions at the final time $T = 7$ as a function of the time step size, with logarithmic scale (left). The error in the norm $L^2\left(0, T;\, L^2(\Omega_t)\right)$ as a function of the time step size, logarithmic scale (right)*

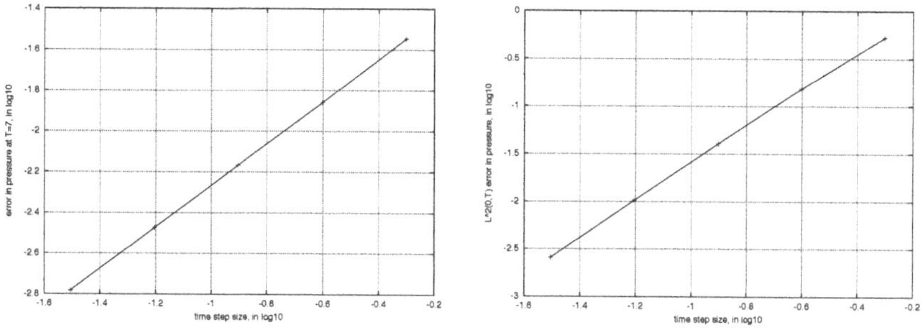

Figure 3.3. *Pressure. The error in the norm $L^2(\Omega_T)$ between the exact and calculated solutions at the final time $T = 7$, as a function of the time step size, with logarithmic scale (left). The error in the norm $L^2\left(0,T;\, L^2(\Omega_t)\right)$ as a function of the time step size, with logarithmic scale (right)*

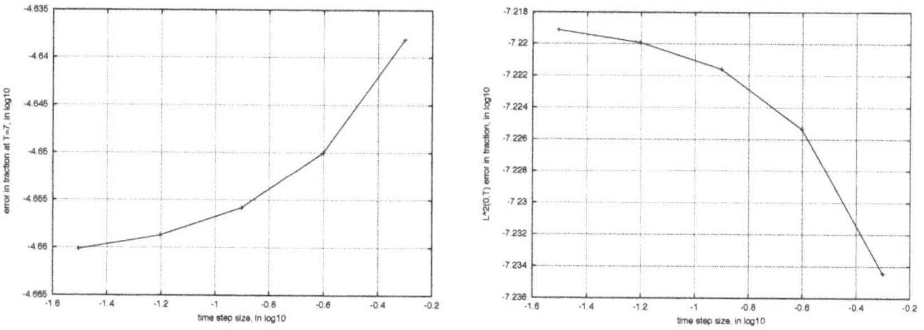

Figure 3.4. *The traction forces on the right-hand boundary. The error in the norm $L^2\left(\mathcal{A}_T(\widehat{\Gamma}_N)\right)$ between the exact and calculated solutions at the final time $T = 7$ as a function of the time step size, with logarithmic scale (left). The error in the norm $L^2\left(0,T;\, L^2\left(\mathcal{A}_t(\widehat{\Gamma}_N)\right)\right)$ as a function of the time step size, with logarithmic scale (right)*

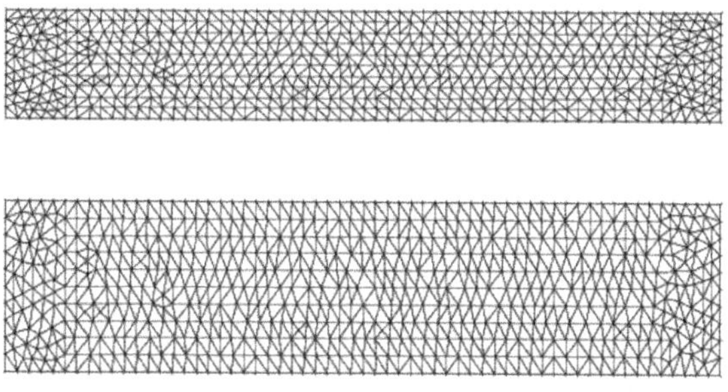

Figure 3.5. *The mesh at time $t = 0$ (above) and time $T = 7$ (below) obtained by harmonic extension*

4

Linear Elastodynamics

4.1. Two-dimensional linear elasticity

4.1.1. *The system of partial differential equations*

Let $\Omega \subset \mathbb{R}^2$ be an open, non-empty, connected, bounded set with Lipschitz boundary $\partial\Omega = \overline{\Gamma}_D \cup \overline{\Gamma}_N$, such that $\Gamma_D \cap \Gamma_N = \emptyset$ and $leng(\Gamma_D) > 0$, $leng(\Gamma_N) > 0$. We are interested in the time evolution over the interval $[0, T]$, where $T > 0$, of an elastic structure that, in its unstressed and non-deformed state, occupies the domain Ω. We denote with

$$\mathbf{U}(X_1, X_2, t) = (U_1(X_1, X_2, t), U_2(X_1, X_2, t))$$

the displacement vector of the general point $\mathbf{X} = (X_1, X_2) \in \Omega$ at time t.

We consider the strain tensor $\epsilon(\mathbf{U})$ given by the relation:

$$\epsilon(\mathbf{U}) = \begin{pmatrix} \epsilon_{11}(\mathbf{U}) & \epsilon_{12}(\mathbf{U}) \\ \epsilon_{21}(\mathbf{U}) & \epsilon_{22}(\mathbf{U}) \end{pmatrix} = \begin{pmatrix} \frac{\partial U_1}{\partial X_1} & \frac{1}{2}\left(\frac{\partial U_1}{\partial X_2} + \frac{\partial U_2}{\partial X_1}\right) \\ \frac{1}{2}\left(\frac{\partial U_2}{\partial X_1} + \frac{\partial U_1}{\partial X_2}\right) & \frac{\partial U_2}{\partial X_2} \end{pmatrix}.$$

Let $E > 0$ be Young's modulus $[Pa]$ and $\nu \in]0, \frac{1}{2}[$ be Poisson's ratio $[dimensionless]$. The Lamé parameters λ and μ are given by the formulae:

$$\lambda = \frac{\nu E}{(1 - 2\nu)(1 + \nu)}, \quad \mu = \frac{E}{2(1 + \nu)}.$$

We consider the stress tensor $\sigma(\mathbf{U})$:

$$\sigma(\mathbf{U}) = \begin{pmatrix} \sigma_{11}(\mathbf{U}) & \sigma_{12}(\mathbf{U}) \\ \sigma_{21}(\mathbf{U}) & \sigma_{22}(\mathbf{U}) \end{pmatrix} = \lambda\, tr(\epsilon(\mathbf{U}))\mathbf{I} + 2\mu\epsilon(\mathbf{U}) =$$

$$\begin{pmatrix} \lambda\,(\epsilon_{11}(\mathbf{U}) + \epsilon_{22}(\mathbf{U})) + 2\mu\epsilon_{11}(\mathbf{U}) & 2\mu\epsilon_{12}(\mathbf{U}) \\ 2\mu\epsilon_{21}(\mathbf{U}) & \lambda\,(\epsilon_{11}(\mathbf{U}) + \epsilon_{22}(\mathbf{U})) + 2\mu\epsilon_{22}(\mathbf{U}) \end{pmatrix}.$$

We are looking to find $\mathbf{U} = (U_1, U_2) : \overline{\Omega} \times [0, T] \to \mathbb{R}^2$, such that:

$$\rho\frac{\partial^2 U_i}{\partial t^2} - \frac{\partial \sigma_{i1}(\mathbf{U})}{\partial X_1} - \frac{\partial \sigma_{i2}(\mathbf{U})}{\partial X_2} = f_i, \text{ in } \Omega\times]0, T[, \ i = 1, 2 \quad [4.1]$$

$$U_i = g_i, \quad \text{on } \Gamma_D\times]0, T[, \ i = 1, 2 \quad [4.2]$$

$$\sigma_{i1}(\mathbf{U})n_1 + \sigma_{i2}(\mathbf{U})n_2 = h_i, \text{ on } \Gamma_N\times]0, T[, \ i = 1, 2 \quad [4.3]$$

$$U_i(X_1, X_2, 0) = U_{0,i}(X_1, X_2), \text{ in } \Omega, \ i = 1, 2 \quad [4.4]$$

$$\frac{\partial U_i}{\partial t}(X_1, X_2, 0) = U_{1,i}(X_1, X_2), \text{ in } \Omega, \ i = 1, 2 \quad [4.5]$$

where

– $\rho > 0$, the mass density, $[Kg/m^3]$;

– $\mathbf{f} = (f_1, f_2) : \Omega\times]0, T[\to \mathbb{R}^2$, the externally applied forces per unit volume (in general gravitational forces), $[N/m^3]$;

– $\mathbf{g} = (g_1, g_2) : \Gamma_D\times]0, T[\to \mathbb{R}^2$, the displacement imposed on Γ_D, $[m]$;

– $\mathbf{h} = (h_1, h_2) : \Gamma_N\times]0, T[\to \mathbb{R}^2$, the external forces per unit surface acting on Γ_N, $[N/m^2]$;

– $\mathbf{U}_0 = (U_{0,1}, U_{0,2}) : \Omega \to \mathbb{R}^2$, the initial displacement, $[m]$;

– $\mathbf{U}_1 = (U_{1,1}, U_{1,2}) : \Omega \to \mathbb{R}^2$, the initial velocity, $[m/s]$;

are given.

4.1.2. *Variational formulation*

For $\mathbf{U} = (U_1, U_2)$, $\mathbf{v} = (v_1, v_2) : \overline{\Omega} \to \mathbb{R}^2$, we note:

$$
\begin{aligned}
a(\mathbf{U}, \mathbf{v}) &= \int_\Omega \sigma(\mathbf{U}) : \nabla \mathbf{v} \, d\mathbf{X} = \int_\Omega \sigma(\mathbf{U}) : \epsilon(\mathbf{v}) \, d\mathbf{X} \\
&= \int_\Omega \lambda(\nabla \cdot \mathbf{U})(\nabla \cdot \mathbf{v}) \, d\mathbf{X} + \int_\Omega 2\mu \, \epsilon(\mathbf{U}) : \epsilon(\mathbf{v}) d\mathbf{X} \\
&= \int_\Omega \lambda \left(\frac{\partial U_1}{\partial X_1} + \frac{\partial U_2}{\partial X_2} \right) \left(\frac{\partial v_1}{\partial X_1} + \frac{\partial v_2}{\partial X_2} \right) d\mathbf{X} \\
&\quad +2\mu \int_\Omega \left(\frac{\partial U_1}{\partial X_1} \frac{\partial v_1}{\partial X_1} + \frac{1}{2} \left(\frac{\partial U_1}{\partial X_2} + \frac{\partial U_2}{\partial X_1} \right) \left(\frac{\partial v_1}{\partial X_2} + \frac{\partial v_2}{\partial X_1} \right) + \frac{\partial U_2}{\partial X_2} \frac{\partial v_2}{\partial X_2} \right) d\mathbf{X}.
\end{aligned}
$$

Find $\mathbf{U} = (U_1, U_2) : \overline{\Omega} \times]0, T[\to \mathbb{R}^2$, $\mathbf{U} = \mathbf{g}$ on $\Gamma_D \times]0, T[$, such that:

$$
\begin{aligned}
&\int_\Omega \rho \left(\frac{\partial^2 U_1}{\partial t^2} v_1 + \frac{\partial^2 U_2}{\partial t^2} v_2 \right) d\mathbf{X} + a(\mathbf{U}, \mathbf{v}) \\
&= \int_\Omega (f_1 v_1 + f_2 v_2) \, d\mathbf{X} + \int_{\Gamma_N} (h_1 v_1 + h_2 v_2) \, ds \qquad [4.6]
\end{aligned}
$$

$\forall v_1 : \overline{\Omega} \to \mathbb{R}$, $v_1|_{\Gamma_D} = 0$ and $\forall v_2 : \overline{\Omega} \to \mathbb{R}$, $v_2|_{\Gamma_D} = 0$.

4.2. Abstract formulation

Let $(H, (\cdot, \cdot)_H, \|\cdot\|_H)$ and $(V, (\cdot, \cdot)_V, \|\cdot\|_V)$ be two Hilbert spaces such that $V \subset H$ and

$$\forall v \in V, \quad \|v\|_H \leq \|v\|_V. \qquad [4.7]$$

We say that the map $a : V \times V \to \mathbb{R}$ is *bilinear* if and only if

$$
\begin{aligned}
&\forall v \in V, \text{ the function } u \to a(u, v) \text{ is linear,} \\
&\forall u \in V, \text{ the function } v \to a(u, v) \text{ is linear.}
\end{aligned}
\qquad [4.8]
$$

We say that the map $a : V \times V \to \mathbb{R}$ is *symmetric* if and only if

$$\forall u, v \in V, \quad a(u, v) = a(v, u). \tag{4.9}$$

We say that the bilinear map $a : V \times V \to \mathbb{R}$ is *continuous* if and only if

$$\exists M > 0, \ \forall u, v \in V, \quad |a(u, v)| \leq M\|u\|_V\|v\|_V. \tag{4.10}$$

We say that the bilinear map $a : V \times V \to \mathbb{R}$ is *elliptic* if and only if

$$\exists \alpha > 0, \ \forall v \in V, \quad \alpha\|v\|_V^2 \leq a(v, v). \tag{4.11}$$

Let $u_0 \in V$, $u_1 \in H$ and $T > 0$. For simplicity, we consider $f \in C^0(]0, T[, H)$. We are looking for $u \in C^0([0, T], V) \cap C^2([0, T], H)$, such that:

$$\left(\frac{d^2u}{dt^2}(t), v\right)_H + a(u(t), v) = (f(t), v)_H, \quad \forall v \in V, \ \forall t \in]0, T[, \tag{4.12}$$

$$u(0) = u_0, \tag{4.13}$$

$$\frac{du}{dt}(0) = u_1 \tag{4.14}$$

where $a : V \times V \to \mathbb{R}$ is a bilinear, symmetric, continuous and elliptic form.

By using $0 \leq \left\|\frac{1}{\sqrt{\gamma}}v - \sqrt{\gamma}w\right\|_H^2$, where $\gamma > 0$, we obtain

$$\forall v, w \in H, \quad (v, w)_H \leq \frac{1}{2\gamma}\|v\|_H^2 + \frac{\gamma}{2}\|w\|_H^2. \tag{4.15}$$

Using the fact that a is bilinear and symmetric, we have the identities:

$$a(v + w, v + w) = a(v, v) + 2a(v, w) + a(w, w), \quad \forall v, w \in V, \tag{4.16}$$

$$a(v + w, v - w) = a(v, v) - a(w, w), \quad \forall v, w \in V. \tag{4.17}$$

LEMMA 4.1.– *[Discrete Gronwall lemma] Consider $\gamma > 0$, $g_0 > 0$, $\Delta t > 0$ and a sequence $(\phi_n)_{n\in\mathbb{N}}$ such that*

$$\phi_0 \leq g_0 \tag{4.18}$$

$$\phi_n \leq g_0 + \gamma\Delta t(\phi_0 + \cdots + \phi_{n-1}), \quad 1 \leq n. \tag{4.19}$$

Hence, $\phi_n \leq g_0e^{\gamma n\Delta t}$, for $1 \leq n$.

DEMONSTRATION 4.1.– We denote $t_n = \gamma n\Delta t$, for $n \in \mathbb{N}$. The function $t \to e^t$ is strictly increasing, thus

$$\int_{t_k}^{t_{k+1}} e^t dt \geq (t_{k+1} - t_k)e^{t_k} = \gamma\Delta t e^{t_k}.$$

This implies that

$$\int_0^{t_{n+1}} e^t dt = \sum_{k=0}^{n} \int_{t_k}^{t_{k+1}} e^t dt \geq \gamma\Delta t\left(e^{t_0} + \cdots + e^{t_n}\right), \quad [4.20]$$

$$\int_0^{t_1} e^t dt \geq \gamma\Delta t. \quad [4.21]$$

Using [4.19] for $n = 1$, [4.18] and [4.21], we have

$$\phi_1 \leq g_0 + \gamma\Delta t\phi_0 \leq g_0 + \gamma\Delta t g_0 \leq g_0 + g_0 \int_0^{t_1} e^t dt$$
$$= g_0 + g_0(e^{t_1} - 1) = g_0 e^{t_1}.$$

We shall demonstrate this by induction. We assume that $\phi_k \leq g_0 e^{\gamma k\Delta t} = g_0 e^{t_k}$ is true for all $k \leq n$ and we shall prove it for $n + 1$. Using [4.19], the inductive hypothesis and [4.20], we have

$$\phi_{n+1} \leq g_0 + \gamma\Delta t\,(\phi_0 + \cdots + \phi_n)$$
$$\leq g_0 + \gamma\Delta t g_0\left(e^{t_0} + \cdots + e^{t_n}\right)$$
$$\leq g_0 + g_0 \int_0^{t_{n+1}} e^t dt = g_0 + g_0(e^{t_{n+1}} - 1) = g_0 e^{t_{n+1}} = g_0 e^{\gamma(n+1)\Delta t}.$$

□

4.3. Backward Euler scheme

Let $\Delta t = \frac{T}{N}$ where $N \in \mathbb{N}^*$. We denote $t_n = n\Delta t$ and $f^n = f(t_n)$ for every $0 \leq n \leq N$. We approximate $u(t_n)$ with u^n and we assume that $u_0, u_1 \in V$. The backward Euler scheme consists of finding $u^{n+1} \in V$ for every $1 \leq n \leq N - 1$ such that:

$$\left(\frac{u^{n+1} - 2u^n + u^{n-1}}{(\Delta t)^2}, v\right)_H + a\left(u^{n+1}, v\right) = \left(f^{n+1}, v\right)_H, \forall v \in V \quad [4.22]$$

$$u^0 = u_0 \quad [4.23]$$

$$u^1 = u_0 + \Delta t\, u_1. \quad [4.24]$$

The problem [4.22] has a unique solution $u^{n+1} \in V$, by the Lax–Milgram Theorem A.9 in the Appendix. It is a scheme of order 1 in Δt.

THEOREM 4.1.– *There exists a constant C, independent of* Δt, *such that*

$$\left\| \frac{u^{n+1} - u^n}{\Delta t} \right\|_H^2 \leq C, \quad a\left(u^{n+1}, u^{n+1}\right) \leq C, \quad 0 \leq n \leq N-1. \qquad [4.25]$$

DEMONSTRATION 4.2.– The proof for the case $f = 0$ is presented in [DAU 88, vol. 9, p. 915]. We shall start with this case.

Taking $v = u^{n+1} - u^n$ in scheme [4.22], we obtain:

$$\frac{1}{(\Delta t)^2}\left(u^{n+1} - 2u^n + u^{n-1}, u^{n+1} - u^n\right)_H + a\left(u^{n+1}, u^{n+1} - u^n\right)$$
$$= \left(f^{n+1}, u^{n+1} - u^n\right)_H. \qquad [4.26]$$

Using the identity $2(v-w,v)_H = \|v\|_H^2 - \|w\|_H^2 + \|v-w\|_H^2$ with $v = u^{n+1} - u^n$ and $w = u^n - u^{n-1}$, we have:

$$\left(u^{n+1} - 2u^n + u^{n-1}, u^{n+1} - u^n\right)_H$$
$$= \left(u^{n+1} - u^n - (u^n - u^{n-1}), u^{n+1} - u^n\right)_H$$
$$= \frac{1}{2}\|u^{n+1} - u^n\|_H^2 - \frac{1}{2}\|u^n - u^{n-1}\|_H^2$$
$$+ \frac{1}{2}\|u^{n+1} - 2u^n + u^{n-1}\|_H^2. \qquad [4.27]$$

Using the bilinearity and symmetry of a, we have a similar identity $2a(v, v-w) = a(v,v) - a(w,w) + a(v-w, v-w)$. Using this identity with $v = u^{n+1}$ and $w = u^n$, we obtain:

$$a\left(u^{n+1}, u^{n+1} - u^n\right) = \frac{1}{2}a\left(u^{n+1}, u^{n+1}\right) - \frac{1}{2}a\left(u^n, u^n\right)$$
$$+ \frac{1}{2}a\left(u^{n+1} - u^n, u^{n+1} - u^n\right). \qquad [4.28]$$

If we note

$$\phi_n = \frac{1}{2}\left\|\frac{u^{n+1}-u^n}{\Delta t}\right\|_H^2 + \frac{1}{2}a\left(u^{n+1},u^{n+1}\right)$$

then we obtain

$$\phi_n + \frac{1}{2}\|u^{n+1} - 2u^n + u^{n-1}\|_H^2 + \frac{1}{2}a\left(u^{n+1}-u^n,u^{n+1}-u^n\right) = \phi_{n-1}$$

and as a consequence $\phi_n \leq \phi_{n-1} \leq \cdots \leq \phi_0$.

Now, we shall study the case $f \neq 0$.

Using inequality [4.15] with $\gamma = \alpha$, we have

$$\left(f^{n+1},u^{n+1}-u^n\right)_H \leq \frac{1}{2\alpha}\left\|f^{n+1}\right\|_H^2 + \frac{\alpha}{2}\left\|u^{n+1}-u^n\right\|_H^2.$$

Taking account of [4.7] and [4.11], we have

$$\frac{\alpha}{2}\left\|u^{n+1}-u^n\right\|_H^2 \leq \frac{\alpha}{2}\left\|u^{n+1}-u^n\right\|_V^2 \leq \frac{1}{2}a\left(u^{n+1}-u^n,u^{n+1}-u^n\right)$$

and, finally, we obtain

$$\left(f^{n+1},u^{n+1}-u^n\right)_H \leq \frac{1}{2\alpha}\left\|f^{n+1}\right\|_H^2 + \frac{1}{2}a\left(u^{n+1}-u^n,u^{n+1}-u^n\right). \qquad [4.29]$$

Taking account of [4.26]–[4.29] and the fact that $0 \leq \|u^{n+1} - 2u^n + u^{n-1}\|_H^2$, we have:

$$\frac{1}{2}\left\|\frac{u^{n+1}-u^n}{\Delta t}\right\|_H^2 - \frac{1}{2}\left\|\frac{u^n-u^{n-1}}{\Delta t}\right\|_H^2 + \frac{1}{2}a\left(u^{n+1},u^{n+1}\right) - \frac{1}{2}a\left(u^n,u^n\right)$$
$$+\frac{1}{2}a\left(u^{n+1}-u^n,u^{n+1}-u^n\right) \leq \frac{1}{2\alpha}\left\|f^{n+1}\right\|_H^2 + \frac{1}{2}a\left(u^{n+1}-u^n,u^{n+1}-u^n\right)$$

thus

$$\frac{1}{2}\left\|\frac{u^{n+1}-u^n}{\Delta t}\right\|_H^2 + \frac{1}{2}a\left(u^{n+1},u^{n+1}\right)$$
$$\leq \frac{1}{2}\left\|\frac{u^n-u^{n-1}}{\Delta t}\right\|_H^2 + \frac{1}{2}a\left(u^n,u^n\right) + \frac{1}{2\alpha}\left\|f^{n+1}\right\|_H^2.$$

The last inequality can be written as:

$$\phi_n \le \phi_{n-1} + \frac{1}{2\alpha}\left\|f^{n+1}\right\|_H^2, \qquad [4.30]$$

and is valid for $n \ge 1$. We remark that $0 \le \phi_n$.

If $2 \le n$, for $1 \le k \le n-1$, we shall use inequality [4.15] with $\gamma = \frac{1}{2\Delta t}$,

$$\begin{aligned}\left(f^{k+1}, u^{k+1} - u^k\right)_H &\le \Delta t \left\|f^{k+1}\right\|_H^2 + \frac{1}{4\Delta t}\left\|u^{k+1} - u^k\right\|_H^2 \\ &= \Delta t \left\|f^{k+1}\right\|_H^2 + \frac{\Delta t}{4}\left\|\frac{u^{k+1} - u^k}{\Delta t}\right\|_H^2. \qquad [4.31]\end{aligned}$$

If we replace n with k in [4.26]–[4.28], by using [4.31] and the fact that $0 \le \|u^{k+1} - 2u^k + u^{k-1}\|_H^2$, $0 \le a\left(u^{k+1} - u^k, u^{k+1} - u^k\right)$, we obtain

$$\begin{aligned}&\phi_k - \phi_{k-1} \\ &= \frac{1}{2}\left\|\frac{u^{k+1} - u^k}{\Delta t}\right\|_H^2 - \frac{1}{2}\left\|\frac{u^k - u^{k-1}}{\Delta t}\right\|_H^2 + \frac{1}{2}a\left(u^{k+1}, u^{k+1}\right) - \frac{1}{2}a\left(u^k, u^k\right) \\ &\le \Delta t \left\|f^{k+1}\right\|_H^2 + \frac{\Delta t}{4}\left\|\frac{u^{k+1} - u^k}{\Delta t}\right\|_H^2.\end{aligned}$$

We shall write the previous inequality for $k = 1, \ldots, n-1$

$$\begin{aligned}\phi_1 - \phi_0 &\le \Delta t \left\|f^2\right\|_H^2 + \frac{\Delta t}{4}\left\|\frac{u^2 - u^1}{\Delta t}\right\|_H^2 \\ &\vdots \\ \phi_{n-1} - \phi_{n-2} &\le \Delta t \left\|f^n\right\|_H^2 + \frac{\Delta t}{4}\left\|\frac{u^n - u^{n-1}}{\Delta t}\right\|_H^2.\end{aligned}$$

By summing this, we obtain:

$$\phi_{n-1} - \phi_0 \le \Delta t \sum_{k=1}^{n-1}\left\|f^{k+1}\right\|_H^2 + \frac{\Delta t}{4}\sum_{k=1}^{n-1}\left\|\frac{u^{k+1} - u^k}{\Delta t}\right\|_H^2$$

and using [4.30], we have

$$\begin{aligned}\phi_n &\le \phi_0 + \frac{1}{2\alpha}\left\|f^{n+1}\right\|_H^2 + \Delta t\sum_{k=1}^{n-1}\left\|f^{k+1}\right\|_H^2 + \frac{\Delta t}{4}\sum_{k=1}^{n-1}\left\|\frac{u^{k+1}-u^k}{\Delta t}\right\|_H^2\\
&\le \phi_0 + \frac{1}{2\alpha}\max_{t\in[0,T]}\|f(t)\|_H^2 + (n-1)\Delta t\max_{t\in[0,T]}\|f(t)\|_H^2 + \frac{\Delta t}{2}\sum_{k=1}^{n-1}\phi_k\\
&\le \phi_0 + \left(\frac{1}{2\alpha}+T\right)\max_{t\in[0,T]}\|f(t)\|_H^2 + \frac{\Delta t}{2}\sum_{k=0}^{n-1}\phi_k.\end{aligned}$$

Thus, for $n \ge 2$,

$$\phi_n \le \phi_0 + \left(\frac{1}{2\alpha}+T\right)\max_{t\in[0,T]}\|f(t)\|_H^2 + \frac{\Delta t}{2}\sum_{k=0}^{n-1}\phi_k.$$

Looking at [4.30], we see that the previous inequality remains valid for $n = 1$ as well.

Using the initial conditions, we have

$$\begin{aligned}\phi_0 &= \frac{1}{2}\left\|\frac{u^1-u^0}{\Delta t}\right\|_H^2 + \frac{1}{2}a\left(u^1,u^1\right)\\
&\le \frac{1}{2}\|u_1\|_H^2 + \frac{1}{2}a(u_0,u_0) + (\Delta t)a(u_0,u_1) + \frac{(\Delta t)^2}{2}a(u_1,u_1)\\
&\le \frac{1}{2}\|u_1\|_H^2 + \frac{1}{2}a(u_0,u_0) + T\,a(u_0,u_1) + \frac{T^2}{2}a(u_1,u_1).\end{aligned}$$

We note

$$\begin{aligned}g_0 &= \frac{1}{2}\|u_1\|_H^2 + \frac{1}{2}a(u_0,u_0) + T\,a(u_0,u_1) + \frac{T^2}{2}a(u_1,u_1)\\
&\quad+\left(\frac{1}{2\alpha}+T\right)\max_{t\in[0,T]}\|f(t)\|_H^2.\end{aligned}$$

As $\phi_0 \leq g_0$, we can apply the discrete Gronwall lemma and we deduce

$$\phi_n \leq g_0 e^{n\Delta t/2} \leq g_0 e^{T/2} = C. \qquad \square$$

4.4. Implicit centered scheme

We assume that $u_0, u_1 \in V$. For every $n \geq 1$, find $u^{n+1} \in V$ such that

$$\left(\frac{u^{n+1} - 2u^n + u^{n-1}}{(\Delta t)^2}, v\right)_H + a\left(\theta u^{n+1} + (1 - 2\theta)\, u^n + \theta u^{n-1}, v\right)$$

$$= (f^n, v)_H\,, \; \forall v \in V \qquad [4.32]$$

$$u^0 = u_0 \qquad [4.33]$$

$$u^1 = u_0 + (\Delta t)u_1 \qquad [4.34]$$

where $\theta \in [0, \frac{1}{2}]$ is a parameter. The problem [4.32] has a unique solution $u^{n+1} \in V$, by the Lax–Milgram Theorem A.2 in the Appendix. This is a scheme of order 2 in Δt.

THEOREM 4.2.– *If $\frac{1}{2} \geq \theta > \frac{1}{4}$, the sequence $(u^n)_{n\in\mathbb{N}}$ defined by the centered scheme satisfies*

$$\left\|\frac{u^{n+1} - u^n}{\Delta t}\right\|_H \leq 2C, \quad a\left(u^n, u^n\right) \leq \frac{2}{4\theta - 1} C, \quad \forall n. \qquad [4.35]$$

If $\theta = \frac{1}{4}$, the sequence $(u^n)_{n\in\mathbb{N}}$ defined by the centered scheme satisfies

$$\left\|\frac{u^{n+1} - u^n}{\Delta t}\right\|_H \leq 2C, \quad a\left(\frac{u^{n+1} + u^n}{2}, \frac{u^{n+1} + u^n}{2}\right) \leq C, \quad \forall n. \qquad [4.36]$$

The constant C is independent of Δt.

DEMONSTRATION 4.3.– The start of the demonstration is based on [DAU 88, vol. 9, p. 919].

Substituting $v = u^{n+1} - u^{n-1}$ into [4.32], we obtain:

$$\left(\frac{u^{n+1} - 2u^n + u^{n-1}}{(\Delta t)^2}, u^{n+1} - u^{n-1}\right)_H$$
$$+ a\left(\theta u^{n+1} + (1 - 2\theta)\, u^n + \theta u^{n-1}, u^{n+1} - u^{n-1}\right)$$
$$= \left(f^n, u^{n+1} - u^{n-1}\right)_H. \qquad [4.37]$$

We shall study the first term in [4.37]. Using the identity $(v + w, v - w)_H = \|v\|_H^2 - \|w\|_H^2$, we have

$$\frac{1}{(\Delta t)^2}\left(u^{n+1} - 2u^n + u^{n-1}, u^{n+1} - u^{n-1}\right)_H$$
$$= \frac{1}{(\Delta t)^2}\left(u^{n+1} - u^n - u^n + u^{n-1}, u^{n+1} - u^n + u^n - u^{n-1}\right)_H$$
$$= \left\|\frac{u^{n+1} - u^n}{\Delta t}\right\|_H^2 - \left\|\frac{u^n - u^{n-1}}{\Delta t}\right\|_H^2. \qquad [4.38]$$

We now turn to the second term in [4.37]. We have

$$a\left(\theta u^{n+1} + (1 - 2\theta)\, u^n + \theta u^{n-1}, u^{n+1} - u^{n-1}\right)$$
$$= \theta a\left(u^{n+1} + u^{n-1}, u^{n+1} - u^{n-1}\right) + (1 - 2\theta)\, a\left(u^n, u^{n+1} - u^{n-1}\right)$$
$$= \theta a\left(u^{n+1}, u^{n+1}\right) - \theta a\left(u^{n-1}, u^{n-1}\right) + (1 - 2\theta)\, a\left(u^n, u^{n+1} - u^{n-1}\right)$$
$$= \theta a\left(u^{n+1}, u^{n+1}\right) + \theta a\left(u^n, u^n\right) + (1 - 2\theta)\, a\left(u^{n+1}, u^n\right)$$
$$- \theta a\left(u^n, u^n\right) - \theta a\left(u^{n-1}, u^{n-1}\right) - (1 - 2\theta)\, a\left(u^n, u^{n-1}\right)$$
$$= X^n - X^{n-1} \qquad [4.39]$$

where

$$X^n = \theta a\left(u^{n+1}, u^{n+1}\right) + \theta a\left(u^n, u^n\right) + (1 - 2\theta)\, a\left(u^{n+1}, u^n\right).$$

We shall rewrite X^n in a different form to see that $X^n \geq 0$ if $\theta \in [1/4, 1/2]$. As a is symmetric, we have

$$a\left(u^{n+1}+u^n, u^{n+1}+u^n\right) = a\left(u^{n+1}, u^{n+1}\right) + 2a\left(u^{n+1}, u^n\right) + a\left(u^n, u^n\right)$$

thus

$$\begin{aligned}
X^n &= \left(\frac{4\theta-1}{2} + \frac{1-2\theta}{2}\right)\left(a\left(u^{n+1}, u^{n+1}\right) + a\left(u^n, u^n\right)\right) \\
&\quad + \left(\frac{1-2\theta}{2}\right) 2a\left(u^{n+1}, u^n\right) \\
&= \left(\frac{4\theta-1}{2}\right)\left(a\left(u^{n+1}, u^{n+1}\right) + a\left(u^n, u^n\right)\right) \\
&\quad + \left(\frac{1-2\theta}{2}\right)\left(a\left(u^{n+1}, u^{n+1}\right) + a\left(u^n, u^n\right) + 2a\left(u^{n+1}, u^n\right)\right) \\
&= \left(\frac{4\theta-1}{2}\right)\left(a\left(u^{n+1}, u^{n+1}\right) + a\left(u^n, u^n\right)\right) \\
&\quad + \left(\frac{1-2\theta}{2}\right) a\left(u^{n+1}+u^n, u^{n+1}+u^n\right).
\end{aligned} \qquad [4.40]$$

We observe that $X^n \geq 0$, if $\theta \in [1/4, 1/2]$.

By substituting [4.38] and [4.39] into [4.37], we deduce that:

$$\left\|\frac{u^{n+1}-u^n}{\Delta t}\right\|_H^2 + X^n - \left\|\frac{u^n-u^{n-1}}{\Delta t}\right\|_H^2 - X^{n-1} = \left(f^n, u^{n+1}-u^{n-1}\right)_H. \qquad [4.41]$$

If $f = 0$, then

$$\left\|\frac{u^{n+1}-u^n}{\Delta t}\right\|_H^2 + X^n = \cdots = \left\|\frac{u^1-u^0}{\Delta t}\right\|_H^2 - X^0.$$

We continue by studying the general case, $f \neq 0$.

We shall use [4.15] with $\gamma = 1$ to expand the right-hand side of [4.41]. For every $1 \leq k \leq n-1$, we have

$$\left(f^k, u^{k+1}-u^{k-1}\right)_H = \Delta t\left(f^k, \frac{u^{k+1}-u^k}{\Delta t} + \frac{u^k-u^{k-1}}{\Delta t}\right)_H$$

$$= \Delta t\left(f^k, \frac{u^{k+1}-u^k}{\Delta t}\right)_H + \Delta t\left(f^k, \frac{u^k-u^{k-1}}{\Delta t}\right)_H$$

$$\leq \frac{\Delta t}{2}\left(\left\|f^k\right\|_H^2 + \left\|\frac{u^{k+1}-u^k}{\Delta t}\right\|_H^2\right) + \frac{\Delta t}{2}\left(\left\|f^k\right\|_H^2 + \left\|\frac{u^k-u^{k-1}}{\Delta t}\right\|_H^2\right)$$

$$= \Delta t\left\|f^k\right\|_H^2 + \frac{\Delta t}{2}\left\|\frac{u^{k+1}-u^k}{\Delta t}\right\|_H^2 + \frac{\Delta t}{2}\left\|\frac{u^k-u^{k-1}}{\Delta t}\right\|_H^2.$$

For the case $k = n$, we use a different expansion

$$\left(f^n, u^{n+1}-u^{n-1}\right)_H = \left(f^n, u^{n+1}-u^n\right)_H + \left(f^n, u^n-u^{n-1}\right)_H$$

$$= \left(\Delta t f^n, \frac{u^{n+1}-u^n}{\Delta t}\right)_H + \Delta t\left(f^n, \frac{u^n-u^{n-1}}{\Delta t}\right)_H$$

$$\leq \frac{1}{2}\left\|\Delta t f^n\right\|_H^2 + \frac{1}{2}\left\|\frac{u^{n+1}-u^n}{\Delta t}\right\|_H^2 + \frac{\Delta t}{2}\left(\left\|f^n\right\|_H^2 + \left\|\frac{u^n-u^{n-1}}{\Delta t}\right\|_H^2\right).$$

We note $\phi_n = \left\|\frac{u^{n+1}-u^n}{\Delta t}\right\|_H^2 + X^n$, and starting from [4.41] and expanding the right-hand side for the case $1 \leq k \leq n-1$ and the case $k = n$, respectively, we obtain

$$\phi_1 - \phi_0 \leq \Delta t\left\|f^1\right\|_H^2 + \frac{\Delta t}{2}\left\|\frac{u^2-u^1}{\Delta t}\right\|_H^2 + \frac{\Delta t}{2}\left\|\frac{u^1-u^0}{\Delta t}\right\|_H^2$$

$$\vdots$$

$$\phi_{n-1} - \phi_{n-2} \leq \Delta t\left\|f^{n-1}\right\|_H^2 + \frac{\Delta t}{2}\left\|\frac{u^n-u^{n-1}}{\Delta t}\right\|_H^2 + \frac{\Delta t}{2}\left\|\frac{u^{n-1}-u^{u-2}}{\Delta t}\right\|_H^2$$

$$\phi_n - \phi_{n-1} \leq \frac{(\Delta t)^2+\Delta t}{2}\left\|f^n\right\|_H^2 + \frac{1}{2}\left\|\frac{u^{n+1}-u^n}{\Delta t}\right\|_H^2 + \frac{\Delta t}{2}\left\|\frac{u^n-u^{n-1}}{\Delta t}\right\|_H^2.$$

Summing leaves us with

$$\phi_n - \phi_0 \leq \Delta t\sum_{k=1}^{n-1}\left\|f^k\right\|_H^2 + \frac{(\Delta t)^2+\Delta t}{2}\left\|f^n\right\|_H^2 + \frac{1}{2}\left\|\frac{u^{n+1}-u^n}{\Delta t}\right\|_H^2$$

$$+\frac{\Delta t}{2}\left\|\frac{u^1-u^0}{\Delta t}\right\|_H^2 + \Delta t\left\|\frac{u^2-u^1}{\Delta t}\right\|_H^2 + \cdots + \Delta t\left\|\frac{u^n-u^{n-1}}{\Delta t}\right\|_H^2$$

$$
\leq \Delta t \sum_{k=1}^{n} \|f^k\|_H^2 + \frac{(\Delta t)^2}{2} \|f^n\|_H^2 + \frac{1}{2} \left\| \frac{u^{n+1} - u^n}{\Delta t} \right\|_H^2
$$

$$
+ \Delta t \left\| \frac{u^1 - u^0}{\Delta t} \right\|_H^2 + \Delta t \left\| \frac{u^2 - u^1}{\Delta t} \right\|_H^2 + \cdots + \Delta t \left\| \frac{u^n - u^{n-1}}{\Delta t} \right\|_H^2 .
$$

As $\phi_n = \left\| \frac{u^{n+1} - u^n}{\Delta t} \right\|_H^2 + X^n$, we have

$$
\frac{1}{2} \left\| \frac{u^{n+1} - u^n}{\Delta t} \right\|_H^2 + X^n \leq \left\| \frac{u^1 - u^0}{\Delta t} \right\|_H^2 + X^0
$$

$$
+ \left(n\Delta t + \frac{(\Delta t)^2}{2} \right) \max_{t \in [0,T]} \|f(t)\|_H^2
$$

$$
+ \Delta t \left(\left\| \frac{u^1 - u^0}{\Delta t} \right\|_H^2 + \cdots + \left\| \frac{u^n - u^{n-1}}{\Delta t} \right\|_H^2 \right).
$$

We note:

$$
\psi_n = \frac{1}{2} \left\| \frac{u^{n+1} - u^n}{\Delta t} \right\|_H^2 + X^n
$$

and

$$
g_0 = \left\| \frac{u^1 - u^0}{\Delta t} \right\|_H^2 + X^0 + \left(T + \frac{T^2}{2} \right) \max_{t \in [0,T]} \|f(t)\|_H^2 .
$$

We have seen that if $\frac{1}{2} \geq \theta \geq \frac{1}{4}$, then $X^n \geq 0$, so

$$
\psi_n \geq \frac{1}{2} \left\| \frac{u^{n+1} - u^n}{\Delta t} \right\|_H^2 .
$$

Finally, we obtain

$$
\psi_n \leq g_0 + 2\Delta t \left(\psi_0 + \cdots + \psi_{n-1} \right)
$$

and applying the discrete Gronwall lemma, since $\psi_0 \leq g_0$, we obtain the result that

$$\psi_n \leq g_0 e^{2T} = C_1.$$

Taking account of [4.40], if $\frac{1}{2} \geq \theta > \frac{1}{4}$, then

$$\frac{1}{2}\left\|\frac{u^{n+1}-u^n}{\Delta t}\right\|_H^2 + \left(\frac{4\theta-1}{2}\right)\left(a\left(u^{n+1},u^{n+1}\right) + a\left(u^n,u^n\right)\right) \leq C_1,$$

and, if $\theta = \frac{1}{4}$, then

$$\frac{1}{2}\left\|\frac{u^{n+1}-u^n}{\Delta t}\right\|_H^2 + \frac{1}{4}a\left(u^{n+1}+u^n, u^{n+1}+u^n\right) \leq C_1.$$

We have not finished, because $\left\|\frac{u^1-u^0}{\Delta t}\right\|_H^2 + X^0$, which is part of g_0, depends on Δt. If we use the initial conditions [4.33], [4.34] and the continuity of a [4.10], we obtain

$$\left\|\frac{u^1-u^0}{\Delta t}\right\|_H^2 + X^0 = \left\|\frac{u^1-u^0}{\Delta t}\right\|_H^2 + \theta a(u^1,u^1) + \theta a(u^0,u^0)$$

$$+(1-2\theta)a(u^1,u^0) \leq \|u_1\|_H^2 + \theta M\,\|u_0 + \Delta t u_1\|_V^2 + \theta a(u_0,u_0)$$

$$+(1-2\theta)M\,\|u_0+\Delta t u_1\|_V\,\|u_0\|_V$$

$$\leq \|u_1\|_H^2 + \theta M\left(\|u_0\|_V + T\,\|u_1\|_V\right)^2 + \theta a(u_0,u_0)$$

$$+(1-2\theta)M\left(\|u_0\|_V + T\,\|u_1\|_V\right)\|u_0\|_V = C_2$$

giving [4.35] and [4.36] with $C = \left(C_2 + \left(T + \frac{T^2}{2}\right)\max_{t\in[0,T]}\|f(t)\|_H^2\right)e^{2T}$. □

4.5. Mid-point scheme

We approximate $u(t_n)$ with u^n, $\frac{du}{dt}(t_n)$ with w^n and we assume that $u_0, u_1 \in V$. The mid-point scheme consists of finding $u^{n+1} \in V$, $w^{n+1} \in V$ for every

$0 \leq n \leq N-1$, such that:

$$\left(\frac{w^{n+1}-w^n}{\Delta t}, v\right)_H + a\left(\frac{u^{n+1}+u^n}{2}, v\right)$$

$$= \left(\frac{f^{n+1}+f^n}{2}, v\right)_H, \ \forall v \in V, \tag{4.42}$$

$$\frac{u^{n+1}-u^n}{\Delta t} = \frac{w^{n+1}+w^n}{2} \tag{4.43}$$

$$u^0 = u_0 \tag{4.44}$$

$$w^0 = u_1. \tag{4.45}$$

Substituting w^{n+1} from [4.43] into [4.42], we obtain a problem with a unique solution $u^{n+1} \in V$ by the Lax–Milgram Theorem A.9 in the Appendix. If $u^{n+1}, u^n, w^n \in V$, from [4.43] we obtain $w^{n+1} \in V$. If we start with $w^0 = u_1 \in V$ and $u^0 = u_0 \in V$, by induction, we obtain $w^{n+1} \in V$, for all n. It is a scheme of order 2 in Δt.

THEOREM 4.3.– *There exists a constant C, independent of Δt, such that*

$$\left\|w^{n+1}\right\|_H^2 \leq 2C, \quad a\left(u^{n+1}, u^{n+1}\right) \leq C, \quad 0 \leq n \leq N-1. \tag{4.46}$$

DEMONSTRATION 4.4.– From [4.43], we have $\Delta t(w^{n+1} + w^n) = 2(u^{n+1} - u^n)$. We set $v = \Delta t(w^{n+1} + w^n) = 2(u^{n+1} - u^n) \in V$ in [4.42] and we obtain

$$\left(\frac{w^{n+1}-w^n}{\Delta t}, \Delta t(w^{n+1}+w^n)\right)_H + a\left(\frac{u^{n+1}+u^n}{2}, 2(u^{n+1}-u^n)\right)$$

$$= \left(\frac{f^{n+1}+f^n}{2}, \Delta t(w^{n+1}+w^n)\right)_H. \tag{4.47}$$

The first term in [4.47] is equal to

$$\left\|w^{n+1}\right\|_H^2 - \left\|w^n\right\|_H^2 \tag{4.48}$$

and the second is equal to

$$a\left(u^{n+1}, u^{n+1}\right) - a\left(u^n, u^n\right). \tag{4.49}$$

If $f = 0$, then

$$\left\|w^{n+1}\right\|_H^2 + a\left(u^{n+1}, u^{n+1}\right) = \left\|w^n\right\|_H^2 + a\left(u^n, u^n\right) = \cdots = \left\|w^0\right\|_H^2 + a\left(u^0, u^0\right).$$

We shall study the case $f \neq 0$.

For the right-hand side of [4.47], we have the expansion:

$$\left(\frac{f^{n+1} + f^n}{2}, \Delta t(w^{n+1} + w^n)\right)_H = \frac{1}{2}\left(\Delta t f^{n+1}, w^{n+1}\right)_H$$
$$+ \frac{1}{2}\left(\Delta t f^n, w^{n+1}\right)_H + \frac{\Delta t}{2}\left(f^{n+1}, w^n\right)_H + \frac{\Delta t}{2}\left(f^n, w^n\right)_H.$$

Using [4.15] with $\gamma = \Delta t$, we have

$$\left(w^{n+1}, \Delta t f^{n+1}\right)_H \leq \frac{1}{2}\left\|w^{n+1}\right\|_H^2 + \frac{(\Delta t)^2}{2}\left\|f^{n+1}\right\|_H^2$$
$$\left(w^{n+1}, \Delta t f^n\right)_H \leq \frac{1}{2}\left\|w^{n+1}\right\|_H^2 + \frac{(\Delta t)^2}{2}\left\|f^n\right\|_H^2$$

and using inequality [4.15] with $\gamma = 1$, we have

$$\left(f^{n+1}, w^n\right)_H \leq \frac{1}{2}\left\|f^{n+1}\right\|_H^2 + \frac{1}{2}\left\|w^n\right\|_H^2$$
$$\left(f^n, w^n\right)_H \leq \frac{1}{2}\left\|f^n\right\|_H^2 + \frac{1}{2}\left\|w^n\right\|_H^2.$$

We thus obtain

$$\left(\frac{f^{n+1} + f^n}{2}, \Delta t(w^{n+1} + w^n)\right)_H \leq \frac{(\Delta t)^2}{4}\left(\left\|f^{n+1}\right\|_H^2 + \left\|f^n\right\|_H^2\right)$$
$$+ \frac{1}{2}\left\|w^{n+1}\right\|_H^2 + \frac{\Delta t}{4}\left(\left\|f^{n+1}\right\|_H^2 + \left\|f^n\right\|_H^2\right) + \frac{\Delta t}{2}\left\|w^n\right\|_H^2. \qquad [4.50]$$

For $0 \leq k \leq n-1$, using inequality [4.15] with $\gamma = 1$, we have

$$\begin{aligned}
&\left(\frac{f^{k+1}+f^k}{2}, \Delta t(w^{k+1}+w^k)\right)_H = \frac{\Delta t}{2}\left(f^{k+1}, w^{k+1}\right)_H + \frac{\Delta t}{2}\left(f^k, w^{k+1}\right)_H \\
&\quad + \frac{\Delta t}{2}\left(f^{k+1}, w^k\right)_H + \frac{\Delta t}{2}\left(f^k, w^k\right)_H \\
&\leq \frac{\Delta t}{4}\left(\|f^{k+1}\|_H^2 + \|w^{k+1}\|_H^2\right) + \frac{\Delta t}{4}\left(\|f^k\|_H^2 + \|w^{k+1}\|_H^2\right) \\
&\quad + \frac{\Delta t}{4}\left(\|f^{k+1}\|_H^2 + \|w^k\|_H^2\right) + \frac{\Delta t}{4}\left(\|f^k\|_H^2 + \|w^k\|_H^2\right) \\
&= \frac{\Delta t}{2}\left(\|f^{k+1}\|_H^2 + \|f^k\|_H^2 + \|w^{k+1}\|_H^2 + \|w^k\|_H^2\right).
\end{aligned} \qquad [4.51]$$

We note that we have used two different expansions for the right-hand side of [4.47], one for $k = n$ and another for $0 \leq k \leq n-1$. Taking account of [4.47]–[4.51], we obtain the inequalities:

$$\begin{aligned}
&\|w^1\|_H^2 - \|w^0\|_H^2 + a\left(u^1, u^1\right) - a\left(u^0, u^0\right) \\
&\leq (\Delta t)\,\|f\|_\infty^2 + \frac{\Delta t}{2}\|w^1\|_H^2 + \frac{\Delta t}{2}\|w^0\|_H^2 \\
&\vdots \\
&\|w^n\|_H^2 - \|w^{n-1}\|_H^2 + a(u^n, u^n) - a\left(u^{n-1}, u^{n-1}\right) \\
&\leq (\Delta t)\,\|f\|_\infty^2 + \frac{\Delta t}{2}\|w^n\|_H^2 + \frac{\Delta t}{2}\|w^{n-1}\|_H^2 \\
&\|w^{n+1}\|_H^2 - \|w^n\|_H^2 + a\left(u^{n+1}, u^{n+1}\right) - a(u^n, u^n) \\
&\leq \frac{(\Delta t)^2}{2}\|f\|_\infty^2 + \frac{1}{2}\|w^{n+1}\|_H^2 + \frac{\Delta t}{2}\|f\|_\infty^2 + \frac{\Delta t}{2}\|w^n\|_H^2
\end{aligned}$$

where $\|f\|_\infty = \max_{t\in[0,T]}\|f(t)\|_H$. By summing, we have:

$$\begin{aligned}
&\|w^{n+1}\|_H^2 - \|w^0\|_H^2 + a\left(u^{n+1}, u^{n+1}\right) - a\left(u^0, u^0\right) \\
&\leq \frac{(\Delta t)^2}{2}\|f\|_\infty^2 + \Delta t\left(n + \frac{1}{2}\right)\|f\|_\infty^2 + \frac{\Delta t}{2}\|w^0\|_H^2 \\
&+\Delta t\|w^1\|_H^2 + \cdots + \Delta t\|w^n\|_H^2 + \frac{1}{2}\|w^{n+1}\|_H^2 .
\end{aligned}$$

As $\Delta t(n+1) \leq T$ and $\Delta t < T$, we deduce:

$$\frac{1}{2}\left\|w^{n+1}\right\|_H^2 + a\left(u^{n+1}, u^{n+1}\right) \leq \left\|w^0\right\|_H^2 + a\left(u^0, u^0\right) + \left(\frac{T^2}{2} + T\right)\|f\|_\infty^2$$
$$+ \Delta t\left(\left\|w^0\right\|_H^2 + \cdots + \left\|w^n\right\|_H^2\right).$$

We note $\phi_{n+1} = \frac{1}{2}\left\|w^{n+1}\right\|_H^2 + a\left(u^{n+1}, u^{n+1}\right)$ and $g_0 = \left\|w^0\right\|_H^2 + a\left(u^0, u^0\right) + \left(\frac{T^2}{2} + T\right)\|f\|_\infty^2$. We have that $\|w^n\|_H^2 \leq 2\phi_n$. We obtain

$$\phi_{n+1} \leq g_0 + 2\Delta t\left(\phi_0 + \cdots + \phi_n\right).$$

We also have that

$$\phi_0 \leq g_0,$$

thus, by applying the discrete Gronwall lemma, we obtain the result that

$$\phi_{n+1} \leq g_0 e^{2T} = C$$

giving [4.46]. □

4.6. Numerical tests

Let Ω be the rectangle $[ABCD]$ with $A(0, -h/2)$, $B(L, -h/2)$, $C(L, h/2)$, $D(0, h/2)$, length $L = 6\ cm$ and height $h = 0.1\ cm$.

We want to solve the linear elastodynamic equations for the following parameters:

– the mass density $\rho = 1.1\ g/cm^3$;

– Young's modulus $E = 750000\ g/(cm\ s^2)$;

– Poisson's ratio $\nu = 0.3$;

– the exterior force per unit volume

$$\mathbf{f} = \left(0,\ \sin(2\pi t)\frac{X_1(L - X_1)(-4\pi^2)}{9} + \sin(2\pi t)\frac{2\mu}{9}\right),$$

– the exterior force per unit surface acting on $\Gamma_N =]AB[\cup]CD[$

$$\mathbf{h} = \left(\mu \sin(2\pi t)\frac{(L-2X_1)}{9},\ 0\right) \text{ on }]CD[,$$

$$\mathbf{h} = \left(-\mu \sin(2\pi t)\frac{(L-2X_1)}{9},\ 0\right) \text{ on }]AB[,$$

– zero displacement $\mathbf{U} = (0,0)$ on $\Gamma_D =]DA[\cup]BC[$;

– the initial conditions $\mathbf{U}_0 = (0,0)$,

$$\mathbf{U}_1 = \left(0,\ \frac{X_1(L-X_1)}{9}(2\pi)\right).$$

The Lamé parameters λ and μ are given by the formulae: $\lambda = \frac{\nu E}{(1-2\nu)(1+\nu)}$ and $\mu = \frac{E}{2(1+\nu)}$.

The numerical parameters are: the time step size $\Delta t = 0.001\ s$, the number of steps in time $N = 1000$ and the finite element $\mathbb{P}_1$. Calculate:

– the error in the norm $L^2(\Omega)$ between the exact solution $\mathbf{U} = \left(0,\ \frac{X_1(L-X_1)}{9}\sin(2\pi t)\right)$ and the calculated solution;

– the vertical displacement of the point $(0, L/2)$;

using the backward Euler scheme, the implicit centered scheme with $\theta = 0.3$ and the mid-point scheme.

The numerical results obtained with a mesh of 364 triangles and 276 vertices are presented in Figure 4.1. There are no significant differences between the three schemes.

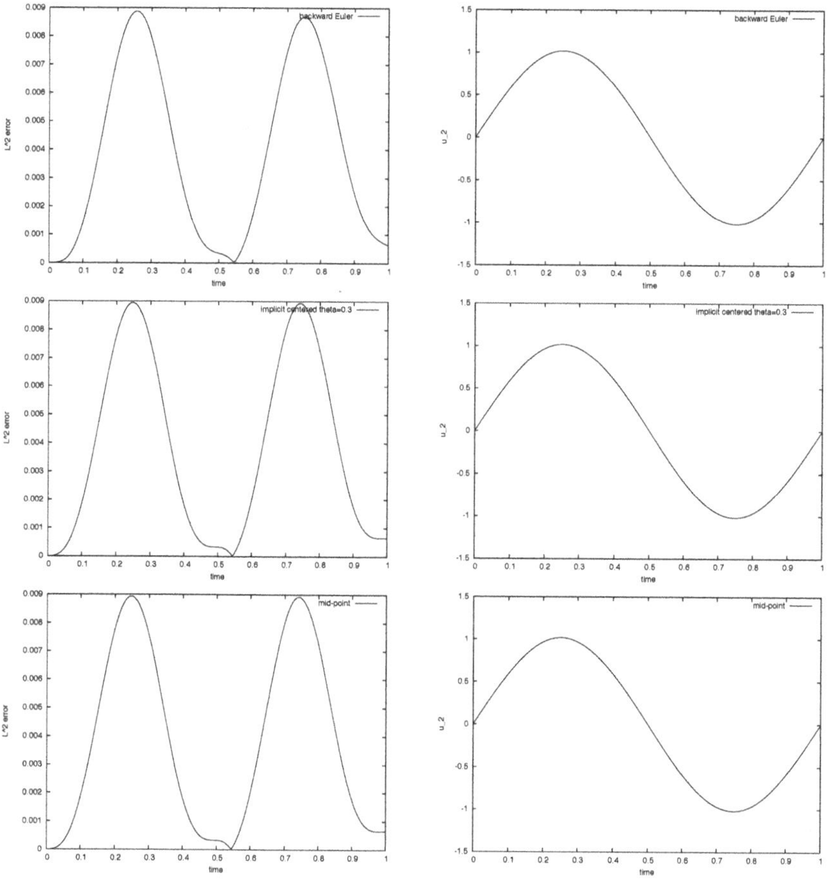

Figure 4.1. *Time evolution of the error* $\|\mathbf{U}^n - \mathbf{U}(\cdot, t_n)\|_{0,\Omega}$ *(left) and the vertical displacement of point* $(0, L/2)$ *(right) for the backward Euler scheme (top), the implicit centered scheme with* $\theta = 0.3$ *(middle) and the mid-point scheme (bottom)*

5

Nonlinear Elastodynamics

5.1. Total Lagrangian formulation

We assume that in its unstressed and non-deformed state, an elastic structure occupies the domain $\Omega \subset \mathbb{R}^2$ with boundary $\partial\Omega = \overline{\Gamma}_D \cup \overline{\Gamma}_N$, such that $\Gamma_D \cap \Gamma_N = \emptyset$ and $leng(\Gamma_D) > 0$, $leng(\Gamma_N) > 0$. We make the assumptions that Ω is an open, non-empty, connected, bounded set with Lipschitz boundary. We also call Ω the reference domain. The structure is fixed on the boundary Γ_D, and we shall apply surface forces on Γ_N.

We assume the existence of a map $\varphi : \overline{\Omega} \times [0, T] \to \mathbb{R}^2$ such that a particle that occupies the position $\mathbf{X} \in \overline{\Omega}$ will occupy the position $\mathbf{x} = \varphi(\mathbf{X}, t)$ in $\overline{\Omega}_t = \varphi\left(\overline{\Omega}, t\right)$. Some authors use the identity $\Omega \equiv \Omega_0 = \varphi(\Omega, 0)$. We shall not use that here, allowing us to set a non-zero initial displacement.

We use the convention $\varphi(\mathbf{X}, t) = \varphi_t(\mathbf{X})$. We call $\mathbf{X} = (X_1, X_2)$ the Lagrangian coordinates and $\mathbf{x} = (x_1, x_2)$ the Eulerian coordinates.

For every $(\mathbf{X}, t) \in \overline{\Omega} \times [0, T]$, we respectively define the displacement, the velocity and the acceleration in Lagrangian coordinates by:

$$\mathbf{U}(\mathbf{X}, t) = \varphi(\mathbf{X}, t) - \mathbf{X},$$

$$\mathbf{V}(\mathbf{X}, t) = \frac{\partial \mathbf{U}}{\partial t}(\mathbf{X}, t),$$

$$\mathbf{A}(\mathbf{X}, t) = \frac{\partial \mathbf{V}}{\partial t}(\mathbf{X}, t) = \frac{\partial^2 \mathbf{U}}{\partial t^2}(\mathbf{X}, t).$$

We assume that, for every $t \in [0, T]$, the map $\varphi_t : \overline{\Omega} \to \overline{\Omega_t}$ is bijective. We can respectively define the displacement, the velocity and the acceleration in Eulerian coordinates by:

$$\mathbf{u}(\mathbf{x}, t) = \mathbf{U}\left(\varphi_t^{-1}(\mathbf{x}), t\right), \quad \mathbf{v}(\mathbf{x}, t) = \mathbf{V}\left(\varphi_t^{-1}(\mathbf{x}), t\right), \quad \mathbf{a}(\mathbf{x}, t) = \mathbf{A}\left(\varphi_t^{-1}(\mathbf{x}), t\right)$$

for every $t \in [0, T]$ and $\mathbf{x} \in \overline{\Omega_t}$.

The gradient, with respect to Eulerian coordinates, of a scalar function $q : \Omega_t \to \mathbb{R}$ and of a vector-valued function $\mathbf{w} = (w_1, w_2) : \Omega_t \to \mathbb{R}^2$ are denoted by

$$\nabla q = \begin{pmatrix} \frac{\partial q}{\partial x_1} \\ \frac{\partial q}{\partial x_2} \end{pmatrix}, \quad \nabla \mathbf{w} = \begin{pmatrix} \frac{\partial w_1}{\partial x_1} & \frac{\partial w_1}{\partial x_2} \\ \frac{\partial w_2}{\partial x_1} & \frac{\partial w_2}{\partial x_2} \end{pmatrix}.$$

The divergence operator with respect to Eulerian coordinates of a vector-valued function $\mathbf{w} = (w_1, w_2) : \Omega_t \to \mathbb{R}^2$ and of a tensor $\boldsymbol{\sigma} = \left(\sigma_{ij}\right)_{1 \le i,j \le 2}$ are denoted by

$$\nabla \cdot \mathbf{w} = \frac{\partial w_1}{\partial x_1} + \frac{\partial w_2}{\partial x_2}, \quad \nabla \cdot \boldsymbol{\sigma} = \begin{pmatrix} \frac{\partial \sigma_{11}}{\partial x_1} + \frac{\partial \sigma_{12}}{\partial x_2} \\ \frac{\partial \sigma_{21}}{\partial x_1} + \frac{\partial \sigma_{22}}{\partial x_2} \end{pmatrix}.$$

Similarly, we use the following notation for the derivatives with respect to the Lagrangian coordinates:

$$\nabla_{\mathbf{X}} \mathbf{U}(\mathbf{X}, t) = \begin{pmatrix} \frac{\partial U_1}{\partial X_1}(\mathbf{X}, t) & \frac{\partial U_1}{\partial X_2}(\mathbf{X}, t) \\ \frac{\partial U_2}{\partial X_1}(\mathbf{X}, t) & \frac{\partial U_2}{\partial X_2}(\mathbf{X}, t) \end{pmatrix},$$

$$\nabla_{\mathbf{X}} \cdot \mathbf{V}(\mathbf{X}, t) = \frac{\partial V_1}{\partial X_1}(\mathbf{X}, t) + \frac{\partial V_2}{\partial X_2}(\mathbf{X}, t).$$

We define the *material derivative* , $\dot{\mathbf{u}}$ or $\frac{D\mathbf{u}}{Dt}$, by

$$\dot{\mathbf{u}}(\mathbf{x}, t) = \frac{D\mathbf{u}}{Dt}(\mathbf{x}, t) = \frac{\partial \mathbf{U}}{\partial t}(\mathbf{X}, t)$$

and we have

$$\mathbf{v} = \dot{\mathbf{u}}, \quad \mathbf{a} = \dot{\mathbf{v}} = \ddot{\mathbf{u}}.$$

If $\mathbf{A}$ is a square matrix, we denote the determinant, the trace, the inverse and the transpose of A with det $\mathbf{A}$, $tr(A)$, $\mathbf{A}^{-1}$ and $\mathbf{A}^T$, respectively. We write $\mathbf{A}^{-T} = \left(\mathbf{A}^{-1}\right)^T$ and cof $\mathbf{A} = (\det \mathbf{A})\left(\mathbf{A}^{-1}\right)^T$ is the cofactor matrix of $\mathbf{A}$.

We use $\mathbf{F}(\mathbf{X}, t) = \mathbf{I} + \nabla_{\mathbf{X}}\mathbf{U}(\mathbf{X}, t)$ to denote the deformation gradient, where $\mathbf{I}$ is the unit matrix and we set $J(\mathbf{X}, t) = \det \mathbf{F}(\mathbf{X}, t)$.

The first and second Piola-Kirchhoff stress tensors are denoted by $\mathbf{\Pi}$ and $\mathbf{\Sigma}$, respectively, and we have the relation $\mathbf{\Pi}(\mathbf{X}, t) = \mathbf{F}(\mathbf{X}, t)\mathbf{\Sigma}(\mathbf{X}, t)$.

The relationship between the Cauchy stress tensor $\boldsymbol{\sigma}$ and the second Piola-Kirchhoff stress tensor $\mathbf{\Sigma}$ is

$$\sigma(\mathbf{x}, t) = \left(\frac{1}{J}\mathbf{F}\mathbf{\Sigma}\mathbf{F}^T\right)(\mathbf{X}, t)$$

where $\mathbf{x} = \mathbf{X} + \mathbf{U}(\mathbf{X}, t)$.

We denote the mass density at time t with $\rho(\mathbf{x}, t)$, and the mass density in the reference domain by $\rho_0(\mathbf{X})$.

Using the Reynolds transport formula, proposition 3.3, for $\frac{d}{dt}\int_{\varphi_t(\omega)} \rho\, d\mathbf{x} = 0$, we obtain the mass conservation equation in Eulerian coordinates

$$\frac{\partial \rho}{\partial t} + \nabla \cdot (\rho\, \mathbf{v}) = 0 \text{ in } \Omega_t,\ t > 0. \quad [5.1]$$

The conservation of momentum in Eulerian coordinates gives:

$$\rho\dot{\mathbf{v}} - \nabla \cdot \boldsymbol{\sigma} = \rho\mathbf{g} \text{ in } \Omega_t,\ t > 0 \quad [5.2]$$

where $\mathbf{g}$ is the gravitational acceleration vector, constant in time and space.

The boundary conditions in the deformed domain are: the displacement condition

$$\varphi_t(\mathbf{X}) = \mathbf{X} \text{ on } \Gamma_D \quad [5.3]$$

and the traction condition in Eulerian coordinates

$$\boldsymbol{\sigma}\mathbf{n} = \boldsymbol{\theta} \text{ on } \varphi_t(\Gamma_N) \quad [5.4]$$

where $\mathbf{n}$ is the unit vector, pointing out of Ω_t, normal to the boundary and $\boldsymbol{\theta}$ are the tractive forces.

Using the fact that $\int_\omega \rho_0(\mathbf{X})\, d\mathbf{X} = \int_{\varphi_t(\omega)} \rho(\mathbf{x},t)\, d\mathbf{x}$, for $t \geq 0$ and every $\omega \subset \Omega$, we obtain

$$\rho(\mathbf{x},t) = \frac{\rho_0(\mathbf{X})}{J(\mathbf{X})}. \tag{5.5}$$

5.1.1. *Strong total Lagrangian formulation*

Find $\mathbf{U} : \overline{\Omega} \times [0,T] \to \mathbb{R}^2$ such that

$$\rho_0(\mathbf{X}) \frac{\partial^2 \mathbf{U}}{\partial t^2}(\mathbf{X},t) - \nabla_{\mathbf{X}} \cdot (\mathbf{F\Sigma})(\mathbf{X},t) = \rho_0(\mathbf{X})\, \mathbf{g},\ \Omega \times]0,T[\tag{5.6}$$

$$\mathbf{U}(\mathbf{X},t) = 0,\ \text{on } \Gamma_D \times]0,T[\tag{5.7}$$

$$(\mathbf{F\Sigma})(\mathbf{X},t)\,\mathbf{N}(\mathbf{X}) = \boldsymbol{\theta}(\varphi_t(\mathbf{X}),t) \left\| J\mathbf{F}^{-T}\mathbf{N} \right\|_{\mathbb{R}^2},\ \text{on } \Gamma_N \times]0,T[\tag{5.8}$$

$$\mathbf{U}(\mathbf{X},0) = \mathbf{U}_0(\mathbf{X}),\ \text{in } \Omega \tag{5.9}$$

$$\frac{\partial \mathbf{U}}{\partial t}(\mathbf{X},0) = \mathbf{U}_1(\mathbf{X}),\ \text{in } \Omega \tag{5.10}$$

where $\mathbf{N}$ is the unit vector pointing out of Ω, normal to the boundary, and $\mathbf{U}_0$ and $\mathbf{U}_1$ are the initial displacement and the initial velocity.

We shall denote by $\mathbf{h}_N = \boldsymbol{\theta}(\varphi_t(\mathbf{X}),t) \left\| J\mathbf{F}^{-T}\mathbf{N} \right\|_{\mathbb{R}^2}$ the tractive forces in the reference domain. We observe that they depend on φ_t, which is an unknown of the problem. In what follows, we shall assume that we know the tractive forces $\mathbf{h}_N : \Gamma_N \times]0,T[\to \mathbb{R}^2$ and the exterior body forces $\mathbf{f} : \Omega \times]0,T[\to \mathbb{R}^2$.

5.1.2. *Weak total Lagrangian formulation*

Find $\mathbf{U} : \overline{\Omega} \times [0,T] \to \mathbb{R}^2$, $\mathbf{U} = 0$ on $\Gamma_D \times]0,T[$ such that

$$\int_\Omega \rho_0 \frac{\partial^2 \mathbf{U}}{\partial t^2} \cdot \mathbf{W}\, d\mathbf{X} + \int_\Omega \mathbf{F\Sigma} : \nabla_{\mathbf{X}} \mathbf{W}\, d\mathbf{X}$$
$$= \int_{\Gamma_N} \mathbf{h}_N \cdot \mathbf{W}\, dS + \int_\Omega \mathbf{f} \cdot \mathbf{W}\, d\mathbf{X} \tag{5.11}$$

for all $\mathbf{W} : \overline{\Omega} \to \mathbb{R}^2$, $\mathbf{W} = 0$ on Γ_D. The initial conditions must be included.

5.2. The St Venant-Kirchhoff model

We note the right Cauchy–Green deformation tensor as $\mathbf{C} = \mathbf{F}^T\mathbf{F}$ and the Green-Lagrange strain tensor as $\mathbf{E} = \frac{1}{2}(\mathbf{C} - \mathbf{I})$. We must provide the expression for the second Piola-Kirchhoff stress tensors $\mathbf{\Sigma}$. A homogeneous material is hyperelastic if there exists a function $\mathcal{W} : \mathcal{M}_{2,2}(\mathbb{R}) \to \mathbb{R}$ such that $\frac{\partial \mathcal{W}(\mathbf{F})}{\partial \mathbf{F}} = \mathbf{\Pi} = \mathbf{F}\mathbf{\Sigma}$, where

$$\frac{\partial \mathcal{W}(\mathbf{F})}{\partial \mathbf{F}} = \begin{pmatrix} \dfrac{\partial \mathcal{W}(\mathbf{F})}{\partial F_{11}} & \dfrac{\partial \mathcal{W}(\mathbf{F})}{\partial F_{12}} \\ \dfrac{\partial \mathcal{W}(\mathbf{F})}{\partial F_{21}} & \dfrac{\partial \mathcal{W}(\mathbf{F})}{\partial F_{22}} \end{pmatrix}.$$

The function $\mathcal{W}$ is known as the strain-energy density of the material.

For the St Venant-Kirchhoff model, we have

$$\mathcal{W} = \frac{\lambda}{2}(tr(\mathbf{E}))^2 + \mu\, tr(\mathbf{E}^2) \qquad [5.12]$$

and

$$\mathbf{\Sigma} = \lambda\, tr(\mathbf{E})\mathbf{I} + 2\mu\mathbf{E} \qquad [5.13]$$

where $\lambda > 0$, $\mu > 0$ are the Lamé parameters. However, $\mathbf{F}$ depends on $\mathbf{U}$ and we can obtain the expression for $\mathbf{E}$ as a function of $\mathbf{U}$:

$$\mathbf{E} = \begin{pmatrix} E_{11} & E_{12} \\ E_{21} & E_{22} \end{pmatrix} = \frac{1}{2}\left((\nabla_{\mathbf{X}}\mathbf{U})^T + \nabla_{\mathbf{X}}\mathbf{U} + (\nabla_{\mathbf{X}}\mathbf{U})^T\nabla_{\mathbf{X}}\mathbf{U}\right).$$

Let $\mathbf{h} : \Omega \to \mathbb{R}^2$ and $\mathbf{h} = (h_1, h_2)$. We can calculate the Fréchet derivatives at the point $\mathbf{U}$ and in the direction $\mathbf{h}$:

$$\frac{dE_{11}}{d\mathbf{U}}(\mathbf{U})\,\mathbf{h} = \frac{1}{2}\left(2\frac{\partial h_1}{\partial x_1} + 2\frac{\partial U_1}{\partial x_1}\frac{\partial h_1}{\partial x_1} + 2\frac{\partial U_1}{\partial x_2}\frac{\partial h_1}{\partial x_2}\right)$$

$$\frac{dE_{22}}{d\mathbf{U}}(\mathbf{U})\,\mathbf{h} = \frac{1}{2}\left(2\frac{\partial h_2}{\partial x_2} + 2\frac{\partial U_2}{\partial x_1}\frac{\partial h_2}{\partial x_1} + 2\frac{\partial U_2}{\partial x_2}\frac{\partial h_2}{\partial x_2}\right)$$

$$\frac{dE_{12}}{d\mathbf{U}}(\mathbf{U})\,\mathbf{h} =$$
$$\frac{1}{2}\left(\frac{\partial h_1}{\partial x_2}+\frac{\partial h_2}{\partial x_1}+\frac{\partial U_2}{\partial x_1}\frac{\partial h_1}{\partial x_1}+\frac{\partial U_2}{\partial x_2}\frac{\partial h_1}{\partial x_2}+\frac{\partial U_1}{\partial x_1}\frac{\partial h_2}{\partial x_1}+\frac{\partial U_1}{\partial x_2}\frac{\partial h_2}{\partial x_2}\right)$$
$$\frac{dE_{21}}{d\mathbf{U}}(\mathbf{U})\,\mathbf{h} = \frac{dE_{12}}{d\mathbf{U}}(\mathbf{U})\,\mathbf{h}$$

or in the matrix form

$$\begin{aligned}\frac{d\mathbf{E}}{d\mathbf{U}}(\mathbf{U})\,\mathbf{h} &= \frac{1}{2}\Big((\nabla_{\mathbf{X}}\mathbf{h})^T + \nabla_{\mathbf{X}}\mathbf{h} + (\nabla_{\mathbf{X}}\mathbf{h})^T\nabla_{\mathbf{X}}\mathbf{U} + (\nabla_{\mathbf{X}}\mathbf{U})^T\nabla_{\mathbf{X}}\mathbf{h}\Big)\\ &= \frac{1}{2}\Big((\nabla_{\mathbf{X}}\mathbf{h})^T\mathbf{F} + \mathbf{F}^T\nabla_{\mathbf{X}}\mathbf{h}\Big).\end{aligned}$$

Let

$$a(\mathbf{U},\mathbf{W}) = \int_\Omega \mathbf{F}(\mathbf{U})\Sigma(\mathbf{U}) : \nabla_{\mathbf{X}}\mathbf{W}\,d\mathbf{X}$$

where we want to calculate $\frac{da}{d\mathbf{U}}(\mathbf{U},\mathbf{W})\mathbf{h}$. We have

$$\begin{aligned}\frac{da}{d\mathbf{U}}(\mathbf{U},\mathbf{W})\mathbf{h} &= \int_\Omega \frac{d\Sigma}{d\mathbf{U}}(\mathbf{U})\mathbf{h} : (\mathbf{F}(\mathbf{U}))^T\nabla_{\mathbf{X}}\mathbf{W}\,d\mathbf{X}\\ &+ \int_\Omega \Sigma(\mathbf{U}) : (\nabla_{\mathbf{X}}\mathbf{h})^T\nabla_{\mathbf{X}}\mathbf{W}\,d\mathbf{X}.\end{aligned}$$

Given that $\Sigma_{ij}(\mathbf{U}) = \lambda(E_{11}(\mathbf{U}) + E_{22}(\mathbf{U}))\,\delta_{ij} + 2\mu E_{ij}(\mathbf{U})$, we can easily deduce $\frac{d\Sigma}{d\mathbf{U}}(\mathbf{U})\mathbf{h}$.

5.3. Total Lagrangian backward Euler scheme

Let $\Delta t = \frac{T}{N}$, where $N \in \mathbb{N}^*$. We denote $t_n = n\Delta t$ and, to simplify the treatment, we assume that the traction $\mathbf{h}_N : \Gamma_N \times [0,T] \to \mathbb{R}^2$ and the body force density $\mathbf{f} : \overline{\Omega} \times [0,T] \to \mathbb{R}^2$ are continuous functions. We set $\mathbf{h}_N^{n+1} = \mathbf{h}_N(\cdot, t_{n+1})$ and $\mathbf{f}^{n+1} = \mathbf{f}(\cdot, t_{n+1})$.

Find $\mathbf{U}^{n+1} : \overline{\Omega} \to \mathbb{R}^2$, $\mathbf{U}^{n+1} = 0$ on Γ_D, such that

$$\int_\Omega \rho_0 \frac{\left(\mathbf{U}^{n+1} - 2\mathbf{U}^n + \mathbf{U}^{n-1}\right)}{(\Delta t)^2} \cdot \mathbf{W} d\mathbf{X} + a\left(\mathbf{U}^{n+1}, \mathbf{W}\right)$$
$$= \int_{\Gamma_N} \mathbf{h}_N^{n+1} \cdot \mathbf{W}\, dS + \int_\Omega \mathbf{f}^{n+1} \cdot \mathbf{W} d\mathbf{X} \qquad [5.14]$$

for all $\mathbf{W} : \overline{\Omega} \to \mathbb{R}^2$, $\mathbf{W} = 0$ on Γ_D. We can impose $\mathbf{U}^0 = \mathbf{U}_0$ and $\mathbf{U}^1 = \mathbf{U}_0 + \Delta t \mathbf{U}_1$. Since the map $\mathbf{U} \to \mathbf{F}(\mathbf{U})\mathbf{\Sigma}(\mathbf{U})$ is nonlinear, we shall use Newton's method to approximate the solution of [5.14].

5.3.1. *Newton's method*

– *Step 0.* Initialization. We set $k = 0$ and $\mathbf{U}^{n+1,0} = \mathbf{U}^n$. We shall construct $\mathbf{U}^{n+1,k}$ for $k = 1, 2, \ldots$

– *Step 1.* Find $\mathbf{h}^k$, the solution of the linear system

$$\int_\Omega \rho_0 \frac{\mathbf{h}^k}{(\Delta t)^2} \cdot \mathbf{W} d\mathbf{X} + \frac{d\,a}{d\,\mathbf{U}}(\mathbf{U}^{n+1,k}, \mathbf{W})\mathbf{h}^k$$
$$+ \int_\Omega \rho_0 \frac{\left(\mathbf{U}^{n+1,k} - 2\mathbf{U}^n + \mathbf{U}^{n-1}\right)}{(\Delta t)^2} \cdot \mathbf{W} d\mathbf{X}$$
$$+a\left(\mathbf{U}^{n+1,k}, \mathbf{W}\right) - \int_{\Gamma_N} \mathbf{h}_N^{n+1} \cdot \mathbf{W}\, dS - \int_\Omega \mathbf{f}^{n+1} \cdot \mathbf{W} d\mathbf{X} = 0 \qquad [5.15]$$

– *Step 2.* If $\mathbf{h}^k$ is small, then we stop.

– *Step 3.* If not, we set $\mathbf{U}^{n+1,k+1} = \mathbf{U}^{n+1,k} + \mathbf{h}^k$; $k \leftarrow k + 1$; go to *Step 1.*

5.4. Total Lagrangian mid-point scheme

Let $\mathbf{f} : \overline{\Omega} \times [0, T] \to \mathbb{R}^2$ be the body force density and assume zero surface tractive forces $\mathbf{h}_N = 0$. To simplify the treatment, we assume that $\mathbf{f}$ is continuous over $\overline{\Omega} \times [0, T]$ and we set $\mathbf{f}^{n+1} = \mathbf{f}(\cdot, t_{n+1})$. Let $\mathbf{U}^n(\mathbf{X})$, $\mathbf{V}^n(\mathbf{X})$ be approximations of $\mathbf{U}(\mathbf{X}, t_n)$, $\mathbf{V}(\mathbf{X}, t_n)$, respectively. We note $\mathbf{F}^n = \mathbf{I} + \nabla_\mathbf{X} \mathbf{U}^n$, $\mathbf{C}^n = (\mathbf{F}^n)^T \mathbf{F}^n$, $\mathbf{E}^n = \frac{1}{2}(\mathbf{C}^n - \mathbf{I})$ and $\mathbf{\Sigma}^n = \lambda\, tr(\mathbf{E}^n)\mathbf{I} + 2\mu \mathbf{E}^n$. We shall also use the notation $\mathbf{F}^{n+1/2} = \frac{\mathbf{F}^{n+1} + \mathbf{F}^n}{2}$ and $\mathbf{\Sigma}^{n+1/2} = \frac{\mathbf{\Sigma}^{n+1} + \mathbf{\Sigma}^n}{2}$.

We can approximate [5.11] by the scheme: find $\mathbf{U}^{n+1}, \mathbf{V}^{n+1} : \overline{\Omega} \to \mathbb{R}^2$, $\mathbf{U}^{n+1}, \mathbf{V}^{n+1} \in C(\overline{\Omega}) \cap C^1(\Omega)$, $\mathbf{U}^{n+1} = 0$ on Γ_D, $\mathbf{V}^{n+1} = 0$ on Γ_D, such that

$$\int_\Omega \rho_0 \frac{\left(\mathbf{V}^{n+1} - \mathbf{V}^n\right)}{\Delta t} \cdot \mathbf{W} d\mathbf{X} + \int_\Omega \mathbf{F}^{n+1/2} \mathbf{\Sigma}^{n+1/2} : \nabla_{\mathbf{X}} \mathbf{W} d\mathbf{X}$$

$$= \int_\Omega \frac{\mathbf{f}^{n+1} + \mathbf{f}^n}{2} \cdot \mathbf{W} d\mathbf{X} \qquad [5.16]$$

$$\frac{\mathbf{U}^{n+1} - \mathbf{U}^n}{\Delta t} = \frac{\mathbf{V}^{n+1} + \mathbf{V}^n}{2} \qquad [5.17]$$

for all $\mathbf{W} : \overline{\Omega} \to \mathbb{R}^2$, $\mathbf{W} \in C(\overline{\Omega}) \cap C^1(\Omega)$, $\mathbf{W} = 0$ on Γ_D. We shall initialize the scheme with $\mathbf{U}^0 = \mathbf{U}_0$ and $\mathbf{V}^0 = \mathbf{U}_1$.

In order to facilitate the calculations, if $\mathbf{E}$ is a matrix from $\mathcal{M}_{2,2}(\mathbb{R})$, then we shall use $[\mathbf{E}]$ to denote the vector from $\mathbb{R}^4$, such that

$$[\mathbf{E}]^T = (E_{11}, E_{22}, E_{12}, E_{21}).$$

Using [5.13] and noting

$$\mathcal{A} = \begin{pmatrix} \lambda + 2\mu & \lambda & 0 & 0 \\ \lambda & \lambda + 2\mu & 0 & 0 \\ 0 & 0 & 2\mu & 0 \\ 0 & 0 & 0 & 2\mu \end{pmatrix}$$

we have $[\mathbf{\Sigma}(\mathbf{E})] = \mathcal{A}[\mathbf{E}]$ and

$$\frac{1}{2}[\mathbf{E}]^T \mathcal{A}[\mathbf{E}] = \frac{1}{2}[\mathbf{E}]^T[\mathbf{\Sigma}(\mathbf{E})] = \frac{1}{2}\left(\lambda(E_{11} + E_{22})^2 + 2\mu \sum_{i,j=1}^{2} E_{ij}^2\right) = \mathcal{W}(\mathbf{E})$$

which is the strain-energy density [5.12], because $E_{21} = E_{12}$. We note $\mathcal{W}^n = \mathcal{W}(\mathbf{E}^n)$ and we observe that $\mathcal{W}^n \geq 0$.

THEOREM 5.1.– *We assume that the mass density is constant, $\rho_0(\mathbf{X}) = \rho_0 > 0$.*

i) If $\mathbf{f} = 0$, then

$$\frac{\rho_0}{2} \|\mathbf{V}^0\|_{0,\Omega}^2 + \int_\Omega \mathcal{W}^0 d\mathbf{X} = \frac{\rho_0}{2} \|\mathbf{V}^{n+1}\|_{0,\Omega}^2 + \int_\Omega \mathcal{W}^{n+1} d\mathbf{X}, \ \forall n \geq 0. \quad [5.18]$$

ii) If $\mathbf{f} \neq 0$, *then there exists a constant* $C > 0$ *independent of* Δt, *such that*

$$\frac{3\rho_0}{8}\left\|\mathbf{V}^{n+1}\right\|_{0,\Omega}^2 + \int_\Omega \mathcal{W}^{n+1} d\mathbf{X} \leq C, \quad 0 \leq n \leq N-1. \qquad [5.19]$$

DEMONSTRATION 5.1.– We set $\mathbf{W} = \mathbf{U}^{n+1} - \mathbf{U}^n$ in [5.16]. According to [5.17], we have $\mathbf{W} = \frac{\Delta t}{2}(\mathbf{V}^{n+1} + \mathbf{V}^n)$.

$$\int_\Omega \rho_0 \frac{\left(\mathbf{V}^{n+1} - \mathbf{V}^n\right)}{\Delta t} \cdot \frac{\Delta t}{2}(\mathbf{V}^{n+1} + \mathbf{V}^n) d\mathbf{X}$$
$$= \frac{\rho_0}{2}\int_\Omega \left(\mathbf{V}^{n+1} - \mathbf{V}^n\right)\cdot\left(\mathbf{V}^{n+1} + \mathbf{V}^n\right) d\mathbf{X} = \frac{\rho_0}{2}\left\|\mathbf{V}^{n+1}\right\|_{0,\Omega}^2 - \frac{\rho_0}{2}\left\|\mathbf{V}^n\right\|_{0,\Omega}^2.$$

W shall study the second term in [5.16]. Using the tensor properties $\mathbf{AB} : \mathbf{C} = \mathbf{B} : \mathbf{A}^T\mathbf{C} = \mathbf{A} : \mathbf{CB}^T$ and $\mathbf{A} : \mathbf{B} = \mathbf{A}^T : \mathbf{B}^T$, we obtain

$$\mathbf{F}^{n+1/2}\mathbf{\Sigma}^{n+1/2} : \nabla_\mathbf{X}(\mathbf{U}^{n+1} - \mathbf{U}^n) = \mathbf{F}^{n+1/2}\mathbf{\Sigma}^{n+1/2} : (\mathbf{F}^{n+1} - \mathbf{F}^n)$$
$$= \mathbf{\Sigma}^{n+1/2} : (\mathbf{F}^{n+1/2})^T(\mathbf{F}^{n+1} - \mathbf{F}^n)$$
$$= \mathbf{\Sigma}^{n+1/2} : \frac{1}{2}(\mathbf{F}^{n+1} + \mathbf{F}^n)^T(\mathbf{F}^{n+1} - \mathbf{F}^n)$$
$$= \mathbf{\Sigma}^{n+1/2} : \frac{1}{2}\left((\mathbf{F}^{n+1})^T\mathbf{F}^{n+1} - (\mathbf{F}^{n+1})^T\mathbf{F}^n + (\mathbf{F}^n)^T\mathbf{F}^{n+1} - (\mathbf{F}^n)^T\mathbf{F}^n\right).$$

As $\mathbf{\Sigma}^{n+1/2}$ is symmetric, we have

$$\mathbf{\Sigma}^{n+1/2} : (\mathbf{F}^n)^T\mathbf{F}^{n+1} = (\mathbf{\Sigma}^{n+1/2})^T : (\mathbf{F}^{n+1})^T\mathbf{F}^n = \mathbf{\Sigma}^{n+1/2} : (\mathbf{F}^{n+1})^T\mathbf{F}^n,$$

thus

$$\mathbf{F}^{n+1/2}\mathbf{\Sigma}^{n+1/2} : \nabla_\mathbf{X}(\mathbf{U}^{n+1} - \mathbf{U}^n)$$
$$= \mathbf{\Sigma}^{n+1/2} : \frac{1}{2}\left((\mathbf{F}^{n+1})^T\mathbf{F}^{n+1} - (\mathbf{F}^n)^T\mathbf{F}^n\right)$$
$$= \mathbf{\Sigma}^{n+1/2} : \left(\mathbf{E}^{n+1} - \mathbf{E}^n\right) = \frac{1}{2}\left(\mathbf{\Sigma}^{n+1} + \mathbf{\Sigma}^n\right) : \left(\mathbf{E}^{n+1} - \mathbf{E}^n\right)$$
$$= \frac{1}{2}\mathbf{\Sigma}^{n+1} : \mathbf{E}^{n+1} - \frac{1}{2}\mathbf{\Sigma}^n : \mathbf{E}^n + \frac{1}{2}\mathbf{\Sigma}^n : \mathbf{E}^{n+1} - \frac{1}{2}\mathbf{\Sigma}^{n+1} : \mathbf{E}^n.$$

Using matrix notation, we have

$$\begin{aligned}\boldsymbol{\Sigma}^n : \mathbf{E}^{n+1} &= [\boldsymbol{\Sigma}^n]^T[\mathbf{E}^{n+1}] = [\mathcal{A}\mathbf{E}^n]^T[\mathbf{E}^{n+1}] = [\mathbf{E}^n]^T\mathcal{A}^T[\mathbf{E}^{n+1}] \\ &= [\mathbf{E}^n]^T\mathcal{A}[\mathbf{E}^{n+1}] = [\mathbf{E}^n]^T[\boldsymbol{\Sigma}^{n+1}] = \mathbf{E}^n : \boldsymbol{\Sigma}^{n+1}.\end{aligned}$$

Finally, we obtain

$$\mathbf{F}^{n+1/2}\boldsymbol{\Sigma}^{n+1/2} : \nabla_{\mathbf{X}}(\mathbf{U}^{n+1} - \mathbf{U}^n) = \frac{1}{2}\boldsymbol{\Sigma}^{n+1} : \mathbf{E}^{n+1} - \frac{1}{2}\boldsymbol{\Sigma}^n : \mathbf{E}^n = \mathcal{W}^{n+1} - \mathcal{W}^n.$$

i) If $\mathbf{f} = 0$ then

$$\frac{\rho_0}{2}\left\|\mathbf{V}^{n+1}\right\|_{0,\Omega}^2 - \frac{\rho_0}{2}\left\|\mathbf{V}^n\right\|_{0,\Omega}^2 + \int_\Omega \mathcal{W}^{n+1}d\mathbf{X} - \int_\Omega \mathcal{W}^n d\mathbf{X} = 0$$

which implies [5.18].

ii) If $\mathbf{f} \neq 0$, then

$$\begin{aligned}&\frac{\rho_0}{2}\left\|\mathbf{V}^{n+1}\right\|_{0,\Omega}^2 - \frac{\rho_0}{2}\left\|\mathbf{V}^n\right\|_{0,\Omega}^2 + \int_\Omega \mathcal{W}^{n+1}d\mathbf{X} - \int_\Omega \mathcal{W}^n d\mathbf{X} \\ &= \int_\Omega \frac{\mathbf{f}^{n+1} + \mathbf{f}^n}{2} \cdot \frac{\Delta t}{2}(\mathbf{V}^{n+1} + \mathbf{V}^n)d\mathbf{X} \\ &= \frac{\Delta t}{4}\int_\Omega (\mathbf{f}^{n+1} + \mathbf{f}^n) \cdot (\mathbf{V}^{n+1} + \mathbf{V}^n)d\mathbf{X}.\end{aligned} \qquad [5.20]$$

We set $M = \max_{\overline{\Omega}\times[0,T]} |\mathbf{f}(\mathbf{x}, t)|$. We shall expand the right-hand side of [5.20] as

$$\begin{aligned}&\frac{\Delta t}{4}\int_\Omega (\mathbf{f}^{n+1} + \mathbf{f}^n) \cdot (\mathbf{V}^{n+1} + \mathbf{V}^n)d\mathbf{X} \\ &= \frac{1}{4}\int_\Omega (\Delta t\mathbf{f}^{n+1} + \Delta t\mathbf{f}^n) \cdot \mathbf{V}^{n+1}d\mathbf{X} + \frac{\Delta t}{4}\int_\Omega (\mathbf{f}^{n+1} + \mathbf{f}^n) \cdot \mathbf{V}^n d\mathbf{X} \\ &\leq \frac{1}{8}\left(\frac{1}{\rho_0}\|\Delta t\mathbf{f}^{n+1} + \Delta t\mathbf{f}^n\|_{0,\Omega}^2 + \rho_0\|\mathbf{V}^{n+1}\|_{0,\Omega}^2\right)\end{aligned}$$

$$
\begin{aligned}
&+\frac{\Delta t}{8}\left(\|\mathbf{f}^{n+1}+\mathbf{f}^n\|_{0,\Omega}^2+\|\mathbf{V}^n\|_{0,\Omega}^2\right)\\
&\leq \frac{\rho_0}{8}\|\mathbf{V}^{n+1}\|_{0,\Omega}^2+\frac{\Delta t}{8}\|\mathbf{V}^n\|_{0,\Omega}^2+\left(\frac{(\Delta t)^2}{8\rho_0}+\frac{\Delta t}{8}\right)\|\mathbf{f}^{n+1}+\mathbf{f}^n\|_{0,\Omega}^2\\
&\leq \frac{\rho_0}{8}\|\mathbf{V}^{n+1}\|_{0,\Omega}^2+\frac{\Delta t}{8}\|\mathbf{V}^n\|_{0,\Omega}^2+\left(\frac{(\Delta t)^2}{8\rho_0}+\frac{\Delta t}{8}\right)M^2.
\end{aligned}
$$

For $k = 0, 1, \ldots, n-1$, we have

$$
\begin{aligned}
&\int_\Omega (\mathbf{f}^{k+1}+\mathbf{f}^k)\cdot(\mathbf{V}^{k+1}+\mathbf{V}^k)d\mathbf{X}\\
&=\int_\Omega \mathbf{f}^{k+1}\cdot\mathbf{V}^{k+1}d\mathbf{X}+\int_\Omega \mathbf{f}^{k+1}\cdot\mathbf{V}^{k}d\mathbf{X}+\int_\Omega \mathbf{f}^{k}\cdot\mathbf{V}^{k+1}d\mathbf{X}+\int_\Omega \mathbf{f}^{k}\cdot\mathbf{V}^{k}d\mathbf{X}\\
&\leq \frac{1}{2}\left(\|\mathbf{f}^{k+1}\|_{0,\Omega}^2+\|\mathbf{V}^{k+1}\|_{0,\Omega}^2+\|\mathbf{f}^{k+1}\|_{0,\Omega}^2+\|\mathbf{V}^{k}\|_{0,\Omega}^2+\|\mathbf{f}^{k}\|_{0,\Omega}^2\right)\\
&+\frac{1}{2}\left(\|\mathbf{V}^{k+1}\|_{0,\Omega}^2+\|\mathbf{f}^{k}\|_{0,\Omega}^2+\|\mathbf{V}^{k}\|_{0,\Omega}^2\right)\leq \|\mathbf{V}^{k+1}\|_{0,\Omega}^2+\|\mathbf{V}^{k}\|_{0,\Omega}^2+2M^2.
\end{aligned}
$$

Inequality [5.20] remains valid if we replace n with k for $k = 0, 1, \ldots, n$. Next, we sum to obtain

$$
\begin{aligned}
&\frac{\rho_0}{2}\left\|\mathbf{V}^{n+1}\right\|_{0,\Omega}^2-\frac{\rho_0}{2}\left\|\mathbf{V}^{0}\right\|_{0,\Omega}^2+\int_\Omega \mathcal{W}^{n+1}d\mathbf{X}-\int_\Omega \mathcal{W}^{0}d\mathbf{X}\\
&\leq \frac{\rho_0}{8}\|\mathbf{V}^{n+1}\|_{0,\Omega}^2+\frac{\Delta t}{8}\|\mathbf{V}^n\|_{0,\Omega}^2+\left(\frac{(\Delta t)^2}{8\rho_0}+\frac{\Delta t}{8}\right)M^2\\
&+\frac{\Delta t}{4}\sum_{k=0}^{n-1}\left(\|\mathbf{V}^{k+1}\|_{0,\Omega}^2+\|\mathbf{V}^{k}\|_{0,\Omega}^2+2M^2\right)\\
&=\frac{\rho_0}{8}\|\mathbf{V}^{n+1}\|_{0,\Omega}^2+\frac{3\Delta t}{8}\|\mathbf{V}^n\|_{0,\Omega}^2+\frac{\Delta t}{2}\sum_{k=1}^{n-1}\|\mathbf{V}^{k}\|_{0,\Omega}^2+\frac{\Delta t}{4}\|\mathbf{V}^0\|_{0,\Omega}^2\\
&+\left(\frac{(\Delta t)^2}{8\rho_0}+\frac{\Delta t}{8}+\frac{(\Delta t)n}{2}\right)M^2\\
&\leq \frac{\rho_0}{8}\|\mathbf{V}^{n+1}\|_{0,\Omega}^2+\frac{\Delta t}{2}\sum_{k=0}^{n}\|\mathbf{V}^{k}\|_{0,\Omega}^2+\left(\frac{(\Delta t)^2}{8\rho_0}+\frac{\Delta t}{8}+\frac{(\Delta t)n}{2}\right)M^2
\end{aligned}
$$

and using the fact that $(\Delta t)n < T$, $\Delta t < T$, we have

$$\frac{3\rho_0}{8}\left\|\mathbf{V}^{n+1}\right\|_{0,\Omega}^2 + \int_\Omega \mathcal{W}^{n+1} d\mathbf{X}$$
$$\leq \frac{\rho_0}{2}\left\|\mathbf{V}^0\right\|_{0,\Omega}^2 + \int_\Omega \mathcal{W}^0 d\mathbf{X} + \frac{\Delta t}{2}\sum_{k=0}^{n} \|\mathbf{V}^k\|_{0,\Omega}^2 + \frac{T^2 + 5T\rho_0}{8\rho_0} M^2.$$

We note

$$\phi_{n+1} = \frac{3\rho_0}{8}\left\|\mathbf{V}^{n+1}\right\|_{0,\Omega}^2 + \int_\Omega \mathcal{W}^{n+1} d\mathbf{X},$$
$$g_0 = \frac{\rho_0}{2}\left\|\mathbf{V}^0\right\|_{0,\Omega}^2 + \int_\Omega \mathcal{W}^0 d\mathbf{X} + \frac{T^2 + 5T\rho_0}{8\rho_0} M^2$$

thus we have

$$\phi_{n+1} \leq g_0 + \frac{\Delta t}{2}\sum_{k=0}^{n} \|\mathbf{V}^k\|_{0,\Omega}^2.$$

However, $\mathcal{W}^k \geq 0$ and consequently $\frac{3\rho_0}{8}\|\mathbf{V}^k\|_{0,\Omega}^2 \leq \phi_k$, which implies

$$\phi_{n+1} \leq g_0 + \frac{\Delta t}{2}\frac{8}{3\rho_0}\sum_{k=0}^{n} \phi_k.$$

As $\phi_0 \leq g_0$, $0 < g_0$, we can apply the discrete Gronwall lemma and we obtain

$$\phi_{n+1} \leq g_0 e^{\frac{4\Delta t}{3\rho_0}(n+1)} \leq C = g_0 e^{\frac{4T}{3\rho_0}}. \qquad \square$$

5.5. Numerical tests

Let Ω be the rectangle $[ABCD]$ with $A(0, -h/2)$, $B(L, -h/2)$, $C(L, h/2)$, $D(0, h/2)$, length $L = 6\ cm$ and height $h = 0.1\ cm$.

We want to solve the time-dependent St Venant-Kirchhoff elasticity equations for the following parameters:

– the mass density $\rho_0 = 1.1\ g/cm^3$;

– Young's modulus $E = 3 \times 10^6\ g/(cm\, s^2)$;

– Poisson's ratio $\nu = 0.3$;

– the exterior forces per unit volume $\mathbf{f} = (0,\ -11000)\ g/(cm^2\, s^2)$;

– the exterior forces per unit surface acting on $\Gamma_N =]AB[\cup]CD[$, $\mathbf{h}_N = (0,\ 0)\ g/(cm\, s^2)$;

– zero displacement $\mathbf{U} = (0, 0)$ on $\Gamma_D =]DA[\cup]BC[$;

– the initial conditions $\mathbf{U}_0 = \mathbf{U}_1 = (0, 0)$.

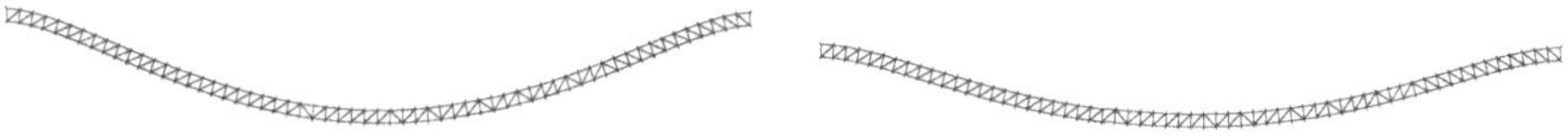

Figure 5.1. *The displacement at t = 0.015 (left) and at t = 0.2 (right)*

The Lamé parameters λ and μ are given by the formulae: $\lambda = \frac{\nu E}{(1-2\nu)(1+\nu)}$ and $\mu = \frac{E}{2(1+\nu)}$.

The numerical parameters are: the time step size $\Delta t = 0.001\ s$, the number of time steps $N = 200$ and the finite element $\mathbb{P}_1$. As a stopping criterion in the Newton method, we can use $\|\mathbf{h}^k\|_{0,\Omega} < 10^{-5}$. Calculate: the deformed mesh, the vertical displacement of the points $(0, L/2)$, $(0, L/4)$ and the volume of the deformed mesh using the total Lagrangian backward Euler scheme.

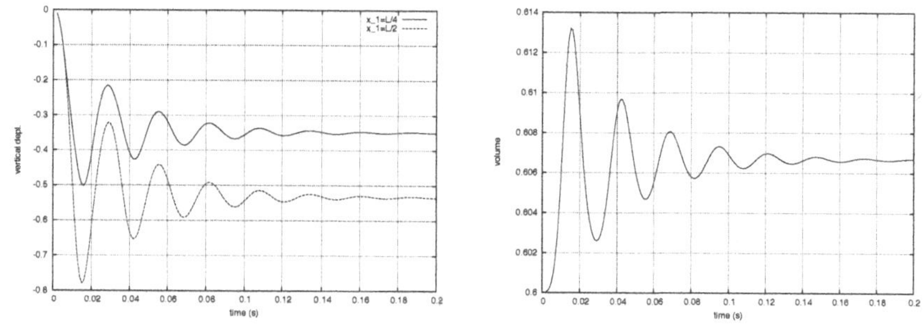

Figure 5.2. *The time evolution of the vertical displacement of points $(0, L/4)$ and $(0, L/2)$ (left) and of the volume (right)*

6

Numerical Schemes for the Fluid-Structure Interaction

6.1. Non-conservative and conservative weak formulations

We assume that, in its unstressed and non-deformed state, an elastic structure occupies the domain Ω^S with boundary $\partial\Omega^S = \overline{\Gamma}_D^S \cup \overline{\Gamma}_N^S \cup \overline{\widehat{\Gamma}}$. The structure is fixed at the boundary Γ_D^S. For the example on the left in Figure 6.1, Ω^S is the rectangle $ABCD$, $\Gamma_D^S =]AB[\cup]CD[$, $\widehat{\Gamma} =]BC[$ and $\Gamma_N^S =]DA[$. Let $\mathbf{U} : \overline{\Omega}^S \times [0,T] \to \mathbb{R}^2$ be the displacement of the structure in Lagrangian coordinates. The point in the non-deformed domain, $\mathbf{X} = (X_1, X_2)^T \in \overline{\Omega}^S$, will occupy the position $\mathbf{x} = \mathbf{X} + \mathbf{U}(\mathbf{X}, t)$ in the deformed domain $\overline{\Omega}_t^S$. We use Γ_t to denote the position of $\widehat{\Gamma}$ at time t after the deformation.

A fluid occupies the domain Ω_t^F with boundary $\partial\Omega_t^F = \overline{\Gamma}_{in}^F \cup \overline{\Gamma}_D^F \cup \overline{\Gamma}_{out}^F \cup \overline{\Gamma}_t$ at time t (see, for example, the right-hand side of Figure 6.1). We set $\Gamma_N^F = \Gamma_{in}^F \cup \Gamma_{out}^F$ and we assume that the boundaries Γ_N^F and Γ_D^F are fixed. The moving boundary Γ_t is called the fluid–structure interface.

For every $t \in [0,T]$, we denote the velocity with $\mathbf{v}(\cdot,t) = \left(v_1^F(\cdot,t), v_2^F(\cdot,t)\right)^T : \overline{\Omega}_t^F \to \mathbb{R}^2$ and the fluid pressure with $p(\cdot,t) : \overline{\Omega}_t^F \to \mathbb{R}$, using Eulerian coordinates, $\mathbf{x} = (x_1, x_2)^T$.

We shall use a linear elasticity model for the structure and we note the stress tensor as

$$\sigma^S = \lambda^S (\nabla_{\mathbf{X}} \cdot \mathbf{U})\mathbf{I} + 2\mu^S \epsilon_{\mathbf{X}}(\mathbf{U})$$

where $\epsilon_{\mathbf{X}}(\mathbf{U}) = \frac{1}{2}\left(\nabla_{\mathbf{X}}\mathbf{U} + (\nabla_{\mathbf{X}}\mathbf{U})^T\right)$ is the linearized strain tensor, $\mathbf{I}$ is the unit (identity) matrix and λ^S, μ^S are the Lamé parameters.

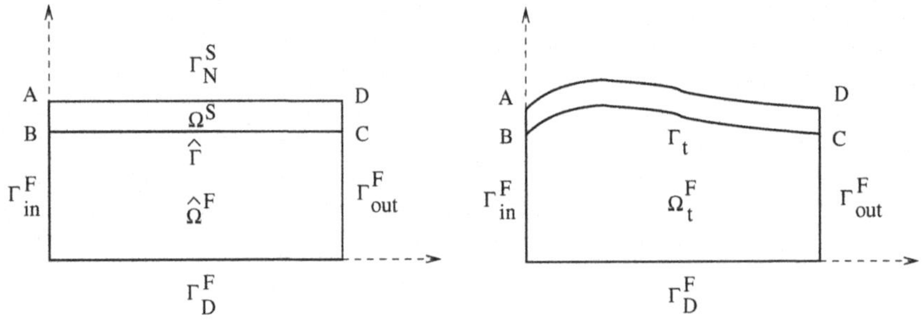

Figure 6.1. *Reference configuration (left) and configuration at time t (right)*

For the fluid, we shall use the Navier–Stokes model and we denote the stress tensor as

$$\sigma^F = -p\mathbf{I} + 2\mu^F \epsilon(\mathbf{v})$$

where $\epsilon(\mathbf{v}) = \frac{1}{2}\left(\nabla\mathbf{v} + (\nabla\mathbf{v})^T\right)$ is the strain rate tensor and μ^F is the dynamic viscosity. To simplify the notation, we use $\nabla\mathbf{v}$ instead of $\nabla_{\mathbf{x}}\mathbf{v}$ to denote the derivatives in Eulerian coordinates.

The fluid–structure interaction problem is as follows: find the structure displacement $\mathbf{U}$, the fluid velocity $\mathbf{v}$ and the fluid pressure p, such that

$$\rho^S \frac{\partial^2 \mathbf{U}}{\partial t^2} - \nabla_{\mathbf{X}} \cdot \sigma^S = \mathbf{f}^S, \text{ in } \Omega^S \times]0,T[\quad [6.1]$$

$$\mathbf{U} = 0, \text{ on } \Gamma_D^S \times]0,T[\quad [6.2]$$

$$\sigma^S \mathbf{N}^S = 0, \text{ on } \Gamma_N^S \times]0,T[\quad [6.3]$$

$$\rho^F \left(\frac{\partial \mathbf{v}}{\partial t} + (\mathbf{v} \cdot \nabla)\mathbf{v}\right) - \nabla \cdot \sigma^F = \mathbf{f}^F, \ \forall t \in]0,T[, \forall \mathbf{x} \in \Omega_t^F \quad [6.4]$$

$$\nabla \cdot \mathbf{v} = 0, \ \forall t \in]0,T[, \forall \mathbf{x} \in \Omega_t^F \quad [6.5]$$

$$\sigma^F \mathbf{n}^F = \mathbf{h}^F, \text{ on } \Gamma_N^F \times]0,T[\quad [6.6]$$

$$\mathbf{v} = 0, \text{ on } \Gamma_D^F \times]0,T[\quad [6.7]$$

$$\mathbf{v}\left(\mathbf{X}+\mathbf{U}\left(\mathbf{X},t\right),t\right) = \frac{\partial \mathbf{U}}{\partial t}\left(\mathbf{X},t\right), \text{ on } \widehat{\Gamma}\times]0,T[\qquad [6.8]$$

$$\left(\sigma^F \mathbf{n}^F\right)_{(\mathbf{X}+\mathbf{U}(\mathbf{X},t),t)} = -\sigma^S\left(\mathbf{X},t\right)\mathbf{N}^S\left(\mathbf{X}\right), \text{ on } \widehat{\Gamma}\times]0,T[\qquad [6.9]$$

$$\mathbf{U}\left(\mathbf{X},0\right) = \mathbf{U}_0\left(\mathbf{X}\right), \text{ in } \Omega^S \qquad [6.10]$$

$$\frac{\partial \mathbf{U}}{\partial t}\left(\mathbf{X},0\right) = \mathbf{U}_1\left(\mathbf{X}\right), \text{ in } \Omega^S \qquad [6.11]$$

$$\mathbf{v}\left(\mathbf{X},0\right) = \mathbf{v}_0\left(\mathbf{X}\right), \text{ in } \Omega_0^F \qquad [6.12]$$

where

– $\rho^S > 0$ and $\rho^F > 0$ are the mass densities of the structure and of the fluid, respectively;

– $\mathbf{f}^S : \Omega^S\times]0,T[\to \mathbb{R}^2$ and $\mathbf{f}^F(\cdot,t) : \Omega_t^F \to \mathbb{R}^2$ are the body forces acting in the structure and in the fluid;

– $\mathbf{N}^S$ is the unit vector pointing out of Ω^S, normal to the boundary $\partial\Omega^S$;

– $\mathbf{h}^F : \Gamma_N^F\times]0,T[\to \mathbb{R}^2$ are the surface forces acting on the fluid, where $\Gamma_N^F = \Gamma_{in}^F \cup \Gamma_{out}^F$;

– $\mathbf{n}^F$ is the unit vector pointing out of Ω_t^F, normal to the boundary $\partial\Omega_t^F$;

– $\mathbf{U}_0$, $\mathbf{U}_1$ are the displacement and the initial velocity of the structure, and $\mathbf{v}_0$ is the initial fluid velocity.

Equations [6.8] and [6.9] are the transmission conditions representing the equality of velocities and the equality of forces at the fluid–structure interface, respectively.

The position of the interface at time t depends on the displacement of the structure, which is an unknown in the system. Hence, a fluid–structure interaction problem is a free-boundary problem.

By multiplying [6.1] by $\mathbf{w}^S : \Omega^S \to \mathbb{R}^2$ such that $\mathbf{w}^S = 0$ on Γ_D^S, integrating over Ω^S and using Green's formula (see theorem A.5), we have

$$\rho^S \int_{\Omega^S} \frac{\partial^2 \mathbf{U}}{\partial t^2} \cdot \mathbf{w}^S \, d\mathbf{X} + \lambda^S \int_{\Omega^S} (\nabla_{\mathbf{X}} \cdot \mathbf{U})(\nabla_{\mathbf{X}} \cdot \mathbf{w}^S) \, d\mathbf{X}$$

$$+2\mu^S \int_{\Omega^S} \epsilon_{\mathbf{X}}(\mathbf{U}) : \epsilon_{\mathbf{X}}\left(\mathbf{w}^S\right) d\mathbf{X}$$

$$= \int_{\Omega^S} \mathbf{f}^S \cdot \mathbf{w}^S \, d\mathbf{X} + \int_{\widehat{\Gamma}} (\sigma^S \mathbf{N}^S) \cdot \mathbf{w}^S \, d\mathbf{S}. \quad [6.13]$$

As the fluid domain is mobile, we shall use the ALE method with the domain $\widehat{\Omega}^F$ and boundary $\partial\widehat{\Omega}^F = \overline{\Gamma}^F_{in} \cup \overline{\Gamma}^F_D \cup \overline{\Gamma}^F_{out} \cup \overline{\widehat{\Gamma}}$ (see, for example, the left-hand side of Figure 6.1) and a map $\mathcal{A}_t : \overline{\widehat{\Omega}}^F \to \overline{\Omega}^F_t$. We saw in Chapter 3 that

$$\left.\frac{\partial \mathbf{v}}{\partial t}\right|_{\widehat{\mathbf{x}}} = \frac{\partial \mathbf{v}}{\partial t} + (\boldsymbol{\vartheta} \cdot \nabla)\, \mathbf{v}.$$

By multiplying [6.4] by $\mathbf{w}^F : \Omega^F_t \to \mathbb{R}^2$ such that $\mathbf{w}^F = 0$ on Γ^F_D, integrating over Ω^F_t and using Green's formula, we have

$$\int_{\Omega^F_t} \rho^F \left.\frac{\partial \mathbf{v}}{\partial t}\right|_{\widehat{\mathbf{x}}} \cdot \mathbf{w}^F \, d\mathbf{x} + \int_{\Omega^F_t} \rho^F [((\mathbf{v} - \boldsymbol{\vartheta}) \cdot \nabla)\mathbf{v}] \cdot \mathbf{w}^F \, d\mathbf{x}$$

$$+ \int_{\Omega^F_t} 2\mu\, \boldsymbol{\epsilon}(\mathbf{v}) : \boldsymbol{\epsilon}\left(\mathbf{w}^F\right) d\mathbf{x} - \int_{\Omega^F_t} \left(\nabla \cdot \mathbf{w}^F\right) p \, d\mathbf{x}$$

$$= \int_{\Omega^F_t} \mathbf{f}^F \cdot \mathbf{w}^F \, d\mathbf{x} + \int_{\Gamma^F_N} \mathbf{h}^F \cdot \mathbf{w}^F ds + \int_{\Gamma_t} (\sigma^F \mathbf{n}^F) \cdot \mathbf{w}^F ds. \quad [6.14]$$

We can construct $\mathcal{A}_t$ such that $\mathcal{A}_t(\mathbf{X}) = \mathbf{X} + \mathbf{U}(\mathbf{X}, t)$, for every $\mathbf{X} \in \widehat{\Gamma}$ and $\mathcal{A}_t(\mathbf{X}) = \mathbf{X}$ for every $\mathbf{X} \in \Gamma^F_{in} \cup \Gamma^F_D \cup \Gamma^F_{out}$. Hence,

$$\boldsymbol{\vartheta}(\mathcal{A}_t(\mathbf{X}), t) = \widehat{\boldsymbol{\vartheta}}(\mathbf{X}, t) = \frac{\partial \mathbf{U}}{\partial t}(\mathbf{X}, t), \quad \forall \mathbf{X} \in \widehat{\Gamma} \quad [6.15]$$

and

$$\boldsymbol{\vartheta}(\mathcal{A}_t(\mathbf{X}), t) = 0, \quad \forall \mathbf{X} \in \Gamma^F_{in} \cup \Gamma^F_D \cup \Gamma^F_{out}. \quad [6.16]$$

Taking account of [6.8], we have that

$$\boldsymbol{\vartheta} = \mathbf{v} \text{ on } \Gamma_t. \quad [6.17]$$

We shall choose $\mathbf{w}^S$ and $\mathbf{w}^F$, such that

$$\mathbf{w}^S(\mathbf{X}) = \mathbf{w}^F(\mathbf{X} + \mathbf{U}(\mathbf{X}, t), t)$$

and in light of [6.9], we can consider that

$$\int_{\widehat{\Gamma}} (\sigma^S \mathbf{N}^S) \cdot \mathbf{w}^S \, d\mathbf{S} + \int_{\Gamma_t} (\sigma^F \mathbf{n}^F) \cdot \mathbf{w}^F ds$$

is negligible.

6.1.1. *Weak non-conservative formulation*

Find:

$$\mathbf{U} \in C^2\left(]0, T[, (L^2(\Omega^S))^2\right) \cap C^1\left([0, T], (H^1(\Omega^S))^2\right)$$
$$\widehat{\mathbf{v}} \in C^1\left(]0, T[, (L^2(\widehat{\Omega}^F))^2\right) \cap C^0\left([0, T], (H^1(\widehat{\Omega}^F))^2\right)$$
$$\widehat{p} \in C^0\left([0, T], L^2(\widehat{\Omega}^F)\right)$$

such that

$$\mathbf{U} = 0, \text{ on } \Gamma_D^S \quad [6.18]$$

$$\mathbf{v} = 0, \text{ on } \Gamma_D^F \quad [6.19]$$

$$\widehat{\mathbf{v}}(\cdot, t) = \frac{\partial \mathbf{U}}{\partial t}(\cdot, t) \text{ on } \widehat{\Gamma} \quad [6.20]$$

and

$$\rho^S \int_{\Omega^S} \frac{\partial^2 \mathbf{U}}{\partial t^2} \cdot \mathbf{w}^S \, d\mathbf{X} + \lambda^S \int_{\Omega^S} (\nabla_{\mathbf{X}} \cdot \mathbf{U})(\nabla_{\mathbf{X}} \cdot \mathbf{w}^S) \, d\mathbf{X}$$
$$+2\mu^S \int_{\Omega^S} \epsilon_{\mathbf{X}}(\mathbf{U}) : \epsilon_{\mathbf{X}}\left(\mathbf{w}^S\right) d\mathbf{X}$$
$$+ \int_{\Omega_t^F} \rho^F \left.\frac{\partial \mathbf{v}}{\partial t}\right|_{\widehat{\mathbf{x}}} \cdot \mathbf{w}^F \, d\mathbf{x} + \int_{\Omega_t^F} \rho^F [((\mathbf{v} - \boldsymbol{\vartheta}) \cdot \nabla)\mathbf{v}] \cdot \mathbf{w}^F \, d\mathbf{x}$$
$$+ \int_{\Omega_t^F} 2\mu^F \boldsymbol{\epsilon}(\mathbf{v}) : \boldsymbol{\epsilon}\left(\mathbf{w}^F\right) d\mathbf{x} - \int_{\Omega_t^F} \left(\nabla \cdot \mathbf{w}^F\right) p \, d\mathbf{x}$$

$$= \int_{\Omega^S} \mathbf{f}^S \cdot \mathbf{w}^S \, d\mathbf{X} + \int_{\Omega_t^F} \mathbf{f}^F \cdot \mathbf{w}^F \, d\mathbf{x} + \int_{\Gamma_N^F} \mathbf{h}^F \cdot \mathbf{w}^F ds,$$

$$\forall \mathbf{w}^S \in (H^1(\Omega^S))^2, \ \mathbf{w}^S = 0 \text{ on } \Gamma_D^S, \ \forall \mathbf{w}^F \in (H^1(\Omega_t^F))^2, \ \mathbf{w}^F = 0 \text{ on } \Gamma_D^F,$$

$$\mathbf{w}^S = \widehat{\mathbf{w}}^F \text{ on } \widehat{\Gamma}, \qquad [6.21]$$

$$-\int_{\Omega_t^F} (\nabla \cdot \mathbf{v}) q \, d\mathbf{x} = 0, \quad \forall q \in L^2(\Omega_t^F), \qquad [6.22]$$

with the initial conditions [6.10]–[6.12].

With regard to the the regularity of the data, we assume that

$$\mathbf{f}^S \in C^0\left([0,T], (L^2(\Omega^S))^2\right),$$

$$\mathbf{h}^F \in C^0\left([0,T], (L^2(\Gamma_N^F))^2\right),$$

$$\mathbf{f}^F(\mathbf{x}, t) = const.$$

It should be noted that the solution of the above system depends neither on the ALE map, nor on the domain velocity. We use the traces on the boundaries, thus we need the domains Ω^S, $\widehat{\Omega}^F$ and Ω_t^F to at least have Lipschitz boundaries.

We can generate test functions on Ω_t^F from test functions on $\widehat{\Omega}^F$. The test function spaces in the ALE domain are:

$$\widehat{W} = \Big\{ \widehat{\mathbf{w}} = (\mathbf{w}^S, \widehat{\mathbf{w}}^F) \in (H^1(\Omega^S))^2 \times (H^1(\widehat{\Omega}^F))^2;$$

$$\mathbf{w}^S = 0 \text{ on } \Gamma_D^S, \ \widehat{\mathbf{w}}^F = 0 \text{ on } \Gamma_D^F, \ \mathbf{w}^S = \widehat{\mathbf{w}}^F \text{ on } \widehat{\Gamma} \Big\},$$

$$\widehat{Q} = L^2(\widehat{\Omega}^F).$$

We shall introduce the test function spaces in Eulerian coordinates

$$W = \Big\{ \mathbf{w}(\cdot, t) = (\mathbf{w}^S(\cdot), \mathbf{w}^F(\cdot, t)); \ \mathbf{w}^S : \Omega^S \to \mathbb{R}^2; \ \mathbf{w}^F(\cdot, t) : \Omega_t^F \to \mathbb{R}^2;$$

$$\exists \widehat{\mathbf{w}} = (\mathbf{w}^S, \widehat{\mathbf{w}}^F) \in \widehat{W}, \ \forall t \in [0,T], \ \forall \mathbf{x} \in \Omega_t^F, \ \mathbf{w}^F(\mathbf{x}, t) = \widehat{\mathbf{w}}^F\left(\mathcal{A}_t^{-1}(\mathbf{x})\right) \Big\},$$

$$Q = \Big\{ q(\cdot, t) : \Omega_t^F \to \mathbb{R}; \ \exists \widehat{q} \in \widehat{Q}, \ \forall t \in [0,T],$$

$$\forall \mathbf{x} \in \Omega_t^F, \ q(\mathbf{x}, t) = \widehat{q}\left(\mathcal{A}_t^{-1}(\mathbf{x})\right) \Big\}.$$

We assume that $\mathcal{A}_t$ is a diffeomorphism and, in this case, if $\mathbf{w} \in W$ and $q \in Q$, then $\mathbf{w}^F(\cdot, t) \in (H^1(\Omega_t^F))^2$ and $q(\cdot, t) \in L^2(\Omega_t^F)$.

As the test function in ALE coordinates does not depend on time (see Chapter 3), we have

$$\int_{\Omega_t^F} \left.\frac{\partial \mathbf{v}}{\partial t}\right|_{\widehat{\mathbf{x}}} \cdot \mathbf{w}^F \, d\mathbf{x} = \frac{d}{dt} \int_{\Omega_t^F} \mathbf{v} \cdot \mathbf{w}^F \, d\mathbf{x} - \int_{\Omega_t^F} (\nabla \cdot \boldsymbol{\vartheta})(\mathbf{v} \cdot \mathbf{w}^F) \, d\mathbf{x}.$$

6.1.2. *Conservative weak formulation*

Find $\mathbf{U}$, $\mathbf{v}$, p with the same regularities as in the non-conservative formulation, such that

$$\begin{aligned}
&\rho^S \int_{\Omega^S} \frac{\partial^2 \mathbf{U}}{\partial t^2} \cdot \mathbf{w}^S \, d\mathbf{X} + \lambda^S \int_{\Omega^S} (\nabla_{\mathbf{X}} \cdot \mathbf{U})(\nabla_{\mathbf{X}} \cdot \mathbf{w}^S) \, d\mathbf{X} \\
&+2\mu^S \int_{\Omega^S} \epsilon_{\mathbf{X}}(\mathbf{U}) : \epsilon_{\mathbf{X}}\left(\mathbf{w}^S\right) d\mathbf{X} \\
&+\rho^F \frac{d}{dt} \int_{\Omega_t^F} \mathbf{v} \cdot \mathbf{w}^F \, d\mathbf{x} - \int_{\Omega_t^F} \rho^F (\nabla \cdot \boldsymbol{\vartheta})(\mathbf{v} \cdot \mathbf{w}^F) \, d\mathbf{x} \\
&+ \int_{\Omega_t^F} \rho^F [((\mathbf{v} - \boldsymbol{\vartheta}) \cdot \nabla)\mathbf{v}] \cdot \mathbf{w}^F \, d\mathbf{x} \\
&+ \int_{\Omega_t^F} 2\mu^F \epsilon(\mathbf{v}) : \epsilon\left(\mathbf{w}^F\right) d\mathbf{x} - \int_{\Omega_t^F} \left(\nabla \cdot \mathbf{w}^F\right) p \, d\mathbf{x} = \int_{\Omega^S} \mathbf{f}^S \cdot \mathbf{w}^S \, d\mathbf{X} \\
&+ \int_{\Omega_t^F} \mathbf{f}^F \cdot \mathbf{w}^F \, d\mathbf{x} + \int_{\Gamma_N^F} \mathbf{h}^F \cdot \mathbf{w}^F ds, \quad \forall \mathbf{w} \in W
\end{aligned} \qquad [6.23]$$

$$- \int_{\Omega_t^F} (\nabla \cdot \mathbf{v}) \, q \, d\mathbf{x} = 0, \quad \forall q \in Q, \qquad [6.24]$$

satisfying [6.18]–[6.20] and the initial conditions [6.10]–[6.12].

In Chapter 3, we introduced $\tilde{c} : (H^1(\Omega_t^F))^2 \times (H^1(\Omega_t^F))^2 \times (H^1(\Omega_t^F))^2 \to \mathbb{R}$

$$\tilde{c}(\mathbf{u}, \mathbf{v}, \mathbf{w}) \overset{déf}{=} \frac{1}{2} \int_{\Omega_t^F} [(\mathbf{u} \cdot \nabla)\mathbf{v}] \cdot \mathbf{w} \, d\mathbf{x} - \frac{1}{2} \int_{\Omega_t^F} [(\mathbf{u} \cdot \nabla)\mathbf{w}] \cdot \mathbf{v} \, d\mathbf{x}$$

which satisfies

$$\tilde{c}(\mathbf{u}, \mathbf{v}, \mathbf{w}) = \int_{\Omega_t^F} [(\mathbf{u} \cdot \nabla)\mathbf{v}] \cdot \mathbf{w}\, d\mathbf{x} + \frac{1}{2} \int_{\Omega_t^F} (\nabla \cdot \mathbf{u})(\mathbf{v} \cdot \mathbf{w})\, d\mathbf{x}$$
$$- \frac{1}{2} \int_{\partial\Omega_t^F} (\mathbf{u} \cdot \mathbf{n})(\mathbf{v} \cdot \mathbf{w})\, ds.$$

We obtain $\nabla \cdot \mathbf{v} = 0$ in Ω_t^F from [6.24] and using [6.16], [6.17] and the fact that $\mathbf{w}^F = 0$ on Γ_D^F, as in Chapter 3, this gives us

$$\int_{\Omega_t^F} [((\mathbf{v} - \boldsymbol{\vartheta}) \cdot \nabla)\mathbf{v}] \cdot \mathbf{w}^F\, d\mathbf{x} = \tilde{c}(\mathbf{v} - \boldsymbol{\vartheta}, \mathbf{v}, \mathbf{w}^F) + \frac{1}{2} \int_{\Omega_t^F} (\nabla \cdot \boldsymbol{\vartheta})(\mathbf{v} \cdot \mathbf{w}^F)\, d\mathbf{x}$$
$$+ \frac{1}{2} \int_{\Gamma_N^F} (\mathbf{v} \cdot \mathbf{n}^F)(\mathbf{v} \cdot \mathbf{w}^F)\, ds.$$

We can obtain an equivalent formulation by substituting the above equation into [6.23]

$$\rho^S \int_{\Omega^S} \frac{\partial^2 \mathbf{U}}{\partial t^2} \cdot \mathbf{w}^S\, d\mathbf{X} + \lambda^S \int_{\Omega^S} (\nabla_{\mathbf{X}} \cdot \mathbf{U})(\nabla_{\mathbf{X}} \cdot \mathbf{w}^S)\, d\mathbf{X}$$
$$+2\mu^S \int_{\Omega^S} \epsilon_{\mathbf{X}}(\mathbf{U}) : \epsilon_{\mathbf{X}}\left(\mathbf{w}^S\right) d\mathbf{X} + \rho^F \frac{d}{dt} \int_{\Omega_t^F} \mathbf{v} \cdot \mathbf{w}^F\, d\mathbf{x}$$
$$-\frac{\rho^F}{2} \int_{\Omega_t^F} (\nabla \cdot \boldsymbol{\vartheta})(\mathbf{v} \cdot \mathbf{w}^F)\, d\mathbf{x} + \rho^F \tilde{c}(\mathbf{v} - \boldsymbol{\vartheta}, \mathbf{v}, \mathbf{w}^F)$$
$$+\frac{\rho^F}{2} \int_{\Gamma_N^F} (\mathbf{v} \cdot \mathbf{n}^F)(\mathbf{v} \cdot \mathbf{w}^F)\, ds + \int_{\Omega_t^F} 2\mu^F\, \epsilon(\mathbf{v}) : \epsilon\left(\mathbf{w}^F\right) d\mathbf{x}$$
$$- \int_{\Omega_t^F} \left(\nabla \cdot \mathbf{w}^F\right) p\, d\mathbf{x} = \int_{\Omega^S} \mathbf{f}^S \cdot \mathbf{w}^S\, d\mathbf{X}$$
$$+ \int_{\Omega_t^F} \mathbf{f}^F \cdot \mathbf{w}^F\, d\mathbf{x} + \int_{\Gamma_N^F} \mathbf{h}^F \cdot \mathbf{w}^F ds, \quad \forall \mathbf{w} \in W. \qquad [6.25]$$

6.2. Discretization in time of the conservative form: implicit domain calculation

In this section, we shall use the formalism presented in [NOB 01] and [QUA 04]. We divide the interval $[0, T]$ into $N \in \mathbb{N}^*$ sub-intervals each of length $\Delta t = T/N$ and we note: $t_n = n\Delta t$, $0 \leq n \leq N$. We shall approximate $\mathbf{U}(\cdot, t_{n+1})$, the displacement of the structure at time t_{n+1}, by $\mathbf{U}^{n+1} : \overline{\Omega}^S \to \mathbb{R}^2$ and we assume that $\mathbf{U}^{n+1}|_{\widehat{\Gamma}}$ is a Lipschitz function.

We construct the interface Γ_{n+1} at time t_{n+1} by

$$\Gamma_{n+1} = (\mathbf{id} + \mathbf{U}^{n+1})(\widehat{\Gamma})$$

where $\mathbf{id}(\mathbf{X}) = \mathbf{X}$ for all $\mathbf{X} \in \mathbb{R}^2$ and the fluid domain Ω^F_{n+1} at time t_{n+1} such that

$$\partial\Omega^F_{n+1} = \overline{\Gamma}^F_D \cup \overline{\Gamma}^F_N \cup \overline{\Gamma}_{n+1}.$$

We shall construct the discrete ALE map $\mathcal{A}_{n+1} : \overline{\widehat{\Omega}}^F \to \overline{\Omega}^F_{n+1}$ by harmonic extension,

$$\mathcal{A}_{n+1}(\widehat{\mathbf{x}}) = \widehat{\mathbf{x}} + \widehat{\mathbf{d}}^{n+1}(\widehat{\mathbf{x}})$$

where

$$\Delta\widehat{\mathbf{d}}^{n+1} = 0, \text{ in } \widehat{\Omega}^F, \qquad [6.26]$$

$$\widehat{\mathbf{d}}^{n+1} = \mathbf{U}^{n+1}, \text{ on } \widehat{\Gamma}, \qquad [6.27]$$

$$\widehat{\mathbf{d}}^{n+1} = 0, \text{ on } \Gamma^F_D \cup \Gamma^F_N. \qquad [6.28]$$

According to remark 3.1, we have at least $\widehat{\mathbf{d}}^{n+1} \in C^2\left(\widehat{\Omega}^F\right) \cap C^0\left(\overline{\widehat{\Omega}}^F\right)$. We assume that $\mathcal{A}_{n+1}$ is a diffeomorphism.

We shall define the discrete domain velocity at time t_{n+1} by

$$\widehat{\boldsymbol{\vartheta}}^{n+1} = \frac{\widehat{\mathbf{d}}^{n+1} - \widehat{\mathbf{d}}^n}{\Delta t}, \quad \boldsymbol{\vartheta}^{n+1} = \widehat{\boldsymbol{\vartheta}}^{n+1} \circ \mathcal{A}^{-1}_{n+1}.$$

We shall use the fluid domain and its velocity at time $t_{n+1/2} = \frac{t_n+t_{n+1}}{2}$ and, for that, we introduce

$$\mathcal{A}_t^{n,n+1}(\widehat{\mathbf{x}}) = \mathcal{A}_n(\widehat{\mathbf{x}})\frac{t_{n+1}-t}{\Delta t} + \mathcal{A}_{n+1}(\widehat{\mathbf{x}})\frac{t-t_n}{\Delta t}$$

for $t \in [t_n, t_{n+1}]$ and $\widehat{\mathbf{x}} \in \overline{\widehat{\Omega}}^F$. We define

$$\mathcal{A}_{n+1/2} = \mathcal{A}_{t_{n+1/2}}^{n,n+1}, \quad \Omega_{n+1/2}^F = \mathcal{A}_{n+1/2}(\widehat{\Omega}^F), \quad \boldsymbol{\vartheta}^{n+1/2} = \widehat{\boldsymbol{\vartheta}}^{n+1} \circ \mathcal{A}_{n+1/2}^{-1}.$$

The test function spaces are constructed from $\widehat{W}$ and $\widehat{Q}$

$$W^{n+1} = \left\{\mathbf{w} = (\mathbf{w}^S, \mathbf{w}^F);\ \exists\widehat{\mathbf{w}} = (\mathbf{w}^S, \widehat{\mathbf{w}}^F) \in \widehat{W},\ \mathbf{w}^F = \widehat{\mathbf{w}}^F \circ \mathcal{A}_{n+1}^{-1}\right\},$$
$$Q^{n+1} = \left\{q : \Omega_{n+1}^F \to \mathbb{R};\ \exists\widehat{q} \in \widehat{Q},\ q = \widehat{q} \circ \mathcal{A}_{n+1}^{-1}\right\}.$$

It should be noted that if $\mathbf{w} \in W^{n+1}$, then $\mathbf{w}^S = 0$ on Γ_D^S, $\mathbf{w}^F = 0$ on Γ_D^F and

$$\mathbf{w}^F(\mathbf{X} + \mathbf{U}^{n+1}(\mathbf{X})) = \mathbf{w}^S(\mathbf{X}), \quad \forall \mathbf{X} \in \widehat{\Gamma}.$$

6.2.1. *Scheme discretized in time for the conservative form*

For $n \geq 0$, find $\mathbf{U}^{n+1} \in (H^1(\Omega^S))^2$, $\mathbf{v}^{n+1} \in (H^1(\Omega_{n+1}^F))^2$, $p^{n+1} \in L^2(\Omega_{n+1}^F)$, such that

$$\mathbf{U}^{n+1} = 0, \text{ on } \Gamma_D^S \quad [6.29]$$

$$\mathbf{v}^{n+1} = 0, \text{ on } \Gamma_D^F \quad [6.30]$$

$$\mathbf{v}^{n+1}(\mathbf{X} + \mathbf{U}^{n+1}(\mathbf{X})) = \frac{\mathbf{U}^{n+1}(\mathbf{X}) - \mathbf{U}^n(\mathbf{X})}{\Delta t}, \quad \forall \mathbf{X} \in \widehat{\Gamma} \quad [6.31]$$

$$\rho^S \int_{\Omega^S} \frac{\mathbf{U}^{n+1} - 2\mathbf{U}^n + \mathbf{U}^{n-1}}{(\Delta t)^2} \cdot \mathbf{w}^S\, d\mathbf{X} + a^S(\mathbf{U}^{n+1}, \mathbf{w}^S)$$

$$+\frac{\rho^F}{\Delta t} \int_{\Omega_{n+1}^F} \mathbf{v}^{n+1} \cdot \mathbf{w}^F d\mathbf{x} - \frac{\rho^F}{\Delta t} \int_{\Omega_n^F} \mathbf{v}^n \cdot \mathbf{w}^F d\mathbf{x}$$

$$-\frac{\rho^F}{2}\int_{\Omega^F_{n+1/2}} (\nabla \cdot \boldsymbol{\vartheta}^{n+1/2})(\mathbf{v}^{n+1} \cdot \mathbf{w}^F)d\mathbf{x}$$

$$+\rho^F \tilde{c}_{n+1}(\mathbf{v}^n - \boldsymbol{\vartheta}^n, \mathbf{v}^{n+1}, \mathbf{w}^F) + \frac{\rho^F}{2}\int_{\Gamma^F_N} (\mathbf{v}^n \cdot \mathbf{n})(\mathbf{v}^{n+1} \cdot \mathbf{w}^F)d\mathbf{s}$$

$$+a^F_{n+1}(\mathbf{v}^{n+1}, \mathbf{w}^F) + b_{n+1}(\mathbf{w}^F, p^{n+1}) = \int_{\Omega^S} \mathbf{f}^{S,n+1} \cdot \mathbf{w}^S \, d\mathbf{X}$$

$$+ \int_{\Omega^F_{n+1}} \mathbf{f}^{F,n+1} \cdot \mathbf{w}^F \, d\mathbf{x} + \int_{\Gamma^F_N} \mathbf{h}^{F,n+1} \cdot \mathbf{w}^F ds, \quad \forall \mathbf{w} \in W^{n+1}, \qquad [6.32]$$

$$b_{n+1}(\mathbf{v}^{n+1}, q) = 0, \quad \forall q \in Q^{n+1}, \qquad [6.33]$$

with the initial conditions

$$\mathbf{U}^{-1} = \mathbf{U}_0 - \Delta t \mathbf{U}_1, \qquad [6.34]$$

$$\mathbf{U}^0 = \mathbf{U}_0, \qquad [6.35]$$

$$\mathbf{v}^0 = \mathbf{v}_0, \qquad [6.36]$$

$$\Delta\widehat{\boldsymbol{\vartheta}}^0 = 0 \text{ in } \widehat{\Omega}^F, \quad \widehat{\boldsymbol{\vartheta}}^0 = \mathbf{U}_1 \text{ on } \widehat{\Gamma}, \quad \widehat{\boldsymbol{\vartheta}}^0 = 0 \text{ on } \Gamma^F_D \cup \Gamma^F_N, \qquad [6.37]$$

where

$$a^S(\mathbf{U}^{n+1}, \mathbf{w}^S) = \lambda^S \int_{\Omega^S} (\nabla_{\mathbf{X}} \cdot \mathbf{U}^{n+1})(\nabla_{\mathbf{X}} \cdot \mathbf{w}^S)\, d\mathbf{X}$$

$$+2\mu^S \int_{\Omega^S} \epsilon_{\mathbf{X}}\left(\mathbf{U}^{n+1}\right) : \epsilon_{\mathbf{X}}\left(\mathbf{w}^S\right) d\mathbf{X},$$

$$a^F_{n+1}(\mathbf{v}^{n+1}, \mathbf{w}^F) = \int_{\Omega^F_{n+1}} 2\mu^F \boldsymbol{\epsilon}\left(\mathbf{v}^{n+1}\right) : \boldsymbol{\epsilon}\left(\mathbf{w}^F\right) d\mathbf{x},$$

$$b_{n+1}(\mathbf{v}^{n+1}, q) = -\int_{\Omega^F_{n+1}} \left(\nabla \cdot \mathbf{v}^{n+1}\right) q \, d\mathbf{x},$$

$$\tilde{c}_{n+1}(\mathbf{u}, \mathbf{v}^{n+1}, \mathbf{w}^F) = \frac{1}{2}\int_{\Omega^F_{n+1}} [(\mathbf{u} \cdot \nabla)\mathbf{v}^{n+1}] \cdot \mathbf{w}^F d\mathbf{x}$$

$$-\frac{1}{2}\int_{\Omega^F_{n+1}} [(\mathbf{u} \cdot \nabla)\mathbf{w}^F] \cdot \mathbf{v}^{n+1} d\mathbf{x}.$$

Using [6.27], [6.31], we have that

$$\mathbf{v}^{n+1}(\mathbf{X} + \mathbf{U}^{n+1}(\mathbf{X})) = \widehat{\boldsymbol{\vartheta}}^{n+1}(\mathbf{X}), \quad \forall \mathbf{X} \in \widehat{\Gamma}$$

which is equivalent to

$$\mathbf{v}^{n+1} = \boldsymbol{\vartheta}^{n+1} \text{ on } \Gamma_{n+1}. \qquad [6.38]$$

As the fluid domain is in motion, we shall work under the hypotheses that the ellipticity constant and the constant in the continuity of the function trace do not depend on the domain motion. More specifically, there exists $\alpha^F > 0$ independent of Ω^F_{n+1}, such that

$$\forall n \in \mathbb{N},\ \forall \mathbf{w}^F \in (H^1(\Omega^F_{n+1}))^2, \quad \alpha^F \left\|\mathbf{w}^F\right\|^2_{1,\Omega^F_{n+1}} \leq a^F_{n+1}(\mathbf{w}^F, \mathbf{w}^F), \qquad [6.39]$$

and there exists $C_{tr} > 0$ independent of Ω^F_{n+1}, such that

$$\forall n \in \mathbb{N},\ \forall \mathbf{w}^F \in (H^1(\Omega^F_{n+1}))^2, \quad \left\|\mathbf{w}^F\right\|_{0,\Gamma^F_N} \leq C_{tr} \left\|\mathbf{w}^F\right\|_{1,\Omega^F_{n+1}}. \qquad [6.40]$$

We note $|\mathbf{v}^{n+1}|^2 = \mathbf{v}^{n+1} \cdot \mathbf{v}^{n+1}$.

THEOREM 6.1.– *We assume that*

$$\int_{\Gamma^F_N} (\mathbf{v}^n \cdot \mathbf{n})|\mathbf{v}^{n+1}|^2 ds \geq 0, \quad 0 \leq n \leq N-1. \qquad [6.41]$$

i) If $\mathbf{f}^S = 0$, $\mathbf{f}^F = 0$, $\mathbf{h}^F = 0$, *then*

$$\frac{\rho^S}{2}\left\|\frac{\mathbf{U}^{n+1} - \mathbf{U}^n}{\Delta t}\right\|^2_{0,\Omega^S} + \frac{1}{2}a^S\left(\mathbf{U}^{n+1}, \mathbf{U}^{n+1}\right) + \frac{\rho^F}{2}\int_{\Omega^F_{n+1}} |\mathbf{v}^{n+1}|^2 d\mathbf{x}$$

$$\leq \frac{\rho^S}{2}\|\mathbf{U}_1\|^2_{0,\Omega^S} + \frac{1}{2}a^S(\mathbf{U}_0, \mathbf{U}_0) + \frac{\rho^F}{2}\int_{\Omega^F_0} |\mathbf{v}_0|^2 d\mathbf{x}. \qquad [6.42]$$

ii) For arbitrary $\mathbf{f}^S \in C^0\left([0,T], (L^2(\Omega^S))^2\right)$, $\mathbf{h}^F \in C^0\left([0,T], (L^2(\Gamma^F_N))^2\right)$, $\mathbf{f}^F(\mathbf{x},t) = const$ *and area*$(\Omega^F_{k+1}) < A_0$, *for* $0 \leq k \leq n$, *hence there exists a* *constant $C > 0$, such that for $0 \leq n \leq N-1$, we have*

$$\frac{\rho^S}{2}\left\|\frac{\mathbf{U}^{n+1}-\mathbf{U}^n}{\Delta t}\right\|^2_{0,\Omega^S} + \frac{1}{2}a^S\left(\mathbf{U}^{n+1},\mathbf{U}^{n+1}\right) + \frac{\rho^F}{2}\int_{\Omega^F_{n+1}} |\mathbf{v}^{n+1}|^2 d\mathbf{x}$$
$$+ \frac{\Delta t}{2}\sum_{k=0}^{n+1} a^F_k(\mathbf{v}^k,\mathbf{v}^k) \leq C.$$

DEMONSTRATION 6.1.– We set $\mathbf{w}^S = \mathbf{U}^{n+1} - \mathbf{U}^n$ and $\mathbf{w}^F = \Delta t \mathbf{v}^{n+1}$. Taking account of [6.31], we have that $\mathbf{w} = (\mathbf{w}^S, \mathbf{w}^F) \in W^{n+1}$ and by substituting it into [6.32], we obtain

$$\rho^S \int_{\Omega^S} \frac{\mathbf{U}^{n+1} - 2\mathbf{U}^n + \mathbf{U}^{n-1}}{(\Delta t)^2}\cdot(\mathbf{U}^{n+1}-\mathbf{U}^n)d\mathbf{X} + a^S(\mathbf{U}^{n+1},\mathbf{U}^{n+1}-\mathbf{U}^n)$$
$$+\rho^F\int_{\Omega^F_{n+1}} \mathbf{v}^{n+1}\cdot\mathbf{v}^{n+1}d\mathbf{x} - \rho^F\int_{\Omega^F_n}\mathbf{v}^n\cdot\mathbf{v}^{n+1}d\mathbf{x}$$
$$-\frac{\Delta t\rho^F}{2}\int_{\Omega^F_{n+1/2}}(\nabla\cdot\boldsymbol{\vartheta}^{n+1/2})(\mathbf{v}^{n+1}\cdot\mathbf{v}^{n+1})d\mathbf{x}$$
$$+\Delta t\rho^F\tilde{c}_{n+1}(\mathbf{v}^n - \boldsymbol{\vartheta}^n,\mathbf{v}^{n+1},\mathbf{v}^{n+1}) + \frac{\Delta t\rho^F}{2}\int_{\Gamma^F_N}(\mathbf{v}^n\cdot\mathbf{n})(\mathbf{v}^{n+1}\cdot\mathbf{v}^{n+1})d\mathbf{s}$$
$$+(\Delta t)a^F_{n+1}(\mathbf{v}^{n+1},\mathbf{v}^{n+1}) + (\Delta t)b_{n+1}(\mathbf{v}^{n+1},p^{n+1})$$
$$=\int_{\Omega^S}\mathbf{f}^{S,n+1}\cdot(\mathbf{U}^{n+1}-\mathbf{U}^n)\,d\mathbf{X} + \Delta t\int_{\Omega^F_{n+1}}\mathbf{f}^{F,n+1}\cdot\mathbf{v}^{n+1}\,d\mathbf{x}$$
$$+\Delta t\int_{\Gamma^F_N}\mathbf{h}^{F,n+1}\cdot\mathbf{v}^{n+1}ds.$$

Using [4.27] and [4.28], we obtain

$$\rho^S \int_{\Omega^S} \frac{\mathbf{U}^{n+1} - 2\mathbf{U}^n + \mathbf{U}^{n-1}}{(\Delta t)^2}\cdot(\mathbf{U}^{n+1}-\mathbf{U}^n)d\mathbf{X} + a^S(\mathbf{U}^{n+1},\mathbf{U}^{n+1}-\mathbf{U}^n)$$
$$\geq \frac{\rho^S}{2}\left\|\frac{\mathbf{U}^{n+1}-\mathbf{U}^n}{\Delta t}\right\|^2_{0,\Omega^S} - \frac{\rho^S}{2}\left\|\frac{\mathbf{U}^{n}-\mathbf{U}^{n-1}}{\Delta t}\right\|^2_{0,\Omega^S} + \frac{1}{2}a^S\left(\mathbf{U}^{n+1},\mathbf{U}^{n+1}\right)$$
$$-\frac{1}{2}a^S(\mathbf{U}^n,\mathbf{U}^n) + \frac{1}{2}a^S\left(\mathbf{U}^{n+1}-\mathbf{U}^n,\mathbf{U}^{n+1}-\mathbf{U}^n\right).$$

Using proposition 3.7, we have

$$\rho^F \int_{\Omega^F_{n+1}} \mathbf{v}^{n+1} \cdot \mathbf{v}^{n+1} d\mathbf{x} - \rho^F \int_{\Omega^F_n} \mathbf{v}^n \cdot \mathbf{v}^{n+1} d\mathbf{x}$$

$$-\frac{\Delta t \rho^F}{2} \int_{\Omega^F_{n+1/2}} (\nabla \cdot \boldsymbol{\vartheta}^{n+1/2})(\mathbf{v}^{n+1} \cdot \mathbf{v}^{n+1}) d\mathbf{x}$$

$$+\Delta t \rho^F \tilde{c}_{n+1}(\mathbf{v}^n - \boldsymbol{\vartheta}^n, \mathbf{v}^{n+1}, \mathbf{v}^{n+1}) + \frac{\Delta t \rho^F}{2} \int_{\Gamma^F_N} (\mathbf{v}^n \cdot \mathbf{n})(\mathbf{v}^{n+1} \cdot \mathbf{v}^{n+1}) ds$$

$$\geq \frac{\rho^F}{2} \int_{\Omega^F_{n+1}} |\mathbf{v}^{n+1}|^2 d\mathbf{x} - \frac{\rho^F}{2} \int_{\Omega^F_n} |\mathbf{v}^n|^2 d\mathbf{x} + \frac{\Delta t \rho^F}{2} \int_{\Gamma^F_N} (\mathbf{v}^n \cdot \mathbf{n}) |\mathbf{v}^{n+1}|^2 ds.$$

Given [6.33], we have $b_{n+1}(\mathbf{v}^{n+1}, p^{n+1}) = 0$. We obtain

$$\frac{\rho^S}{2} \left\| \frac{\mathbf{U}^{n+1} - \mathbf{U}^n}{\Delta t} \right\|^2_{0,\Omega^S} - \frac{\rho^S}{2} \left\| \frac{\mathbf{U}^n - \mathbf{U}^{n-1}}{\Delta t} \right\|^2_{0,\Omega^S} + \frac{1}{2} a^S \left(\mathbf{U}^{n+1}, \mathbf{U}^{n+1} \right)$$

$$-\frac{1}{2} a^S \left(\mathbf{U}^n, \mathbf{U}^n \right) + \frac{1}{2} a^S \left(\mathbf{U}^{n+1} - \mathbf{U}^n, \mathbf{U}^{n+1} - \mathbf{U}^n \right)$$

$$+\frac{\rho^F}{2} \int_{\Omega^F_{n+1}} |\mathbf{v}^{n+1}|^2 d\mathbf{x} - \frac{\rho^F}{2} \int_{\Omega^F_n} |\mathbf{v}^n|^2 d\mathbf{x} + \frac{\Delta t \rho^F}{2} \int_{\Gamma^F_N} (\mathbf{v}^n \cdot \mathbf{n}) |\mathbf{v}^{n+1}|^2 ds$$

$$+(\Delta t) a^F_{n+1}(\mathbf{v}^{n+1}, \mathbf{v}^{n+1})$$

$$\leq \int_{\Omega^S} \mathbf{f}^{S,n+1} \cdot (\mathbf{U}^{n+1} - \mathbf{U}^n) d\mathbf{X} + \Delta t \int_{\Omega^F_{n+1}} \mathbf{f}^{F,n+1} \cdot \mathbf{v}^{n+1} d\mathbf{x}$$

$$+\Delta t \int_{\Gamma^F_N} \mathbf{h}^{F,n+1} \cdot \mathbf{v}^{n+1} ds. \qquad [6.43]$$

i) If $\mathbf{f}^S = 0, \mathbf{f}^F = 0, \mathbf{h}^F = 0$ and $\int_{\Gamma^F_N} (\mathbf{v}^n \cdot \mathbf{n}) |\mathbf{v}^{n+1}|^2 ds \geq 0$, then

$$\frac{\rho^S}{2} \left\| \frac{\mathbf{U}^{n+1} - \mathbf{U}^n}{\Delta t} \right\|^2_{0,\Omega^S} - \frac{\rho^S}{2} \left\| \frac{\mathbf{U}^n - \mathbf{U}^{n-1}}{\Delta t} \right\|^2_{0,\Omega^S} + \frac{1}{2} a^S \left(\mathbf{U}^{n+1}, \mathbf{U}^{n+1} \right)$$

$$-\frac{1}{2} a^S \left(\mathbf{U}^n, \mathbf{U}^n \right) + \frac{\rho^F}{2} \int_{\Omega^F_{n+1}} |\mathbf{v}^{n+1}|^2 d\mathbf{x} - \frac{\rho^F}{2} \int_{\Omega^F_n} |\mathbf{v}^n|^2 d\mathbf{x} \leq 0. \qquad [6.44]$$

Taking account of the initial conditions [6.34]–[6.36], we have [6.42].

ii) In the general case, we shall expand the right-hand side of [6.43].

For the structure, we have similar inequalities to [4.29] and [4.31]

$$\left(\mathbf{f}^{S,n+1}, \mathbf{U}^{n+1} - \mathbf{U}^{n}\right)_{0,\Omega^S} \leq \frac{1}{2\alpha^S} \left\|\mathbf{f}^{S,n+1}\right\|^2_{0,\Omega^S} + \frac{1}{2} a^S\left(\mathbf{U}^{n+1} - \mathbf{U}^{n}, \mathbf{U}^{n+1} - \mathbf{U}^{n}\right) \quad [6.45]$$

and, for $0 \leq k \leq n-1$,

$$\left(\mathbf{f}^{S,k+1}, \mathbf{U}^{k+1} - \mathbf{U}^{k}\right)_{0,\Omega^S} \leq \Delta t \left\|\mathbf{f}^{S,k+1}\right\|^2_{0,\Omega^S} + \frac{1}{4\Delta t} \left\|\mathbf{U}^{k+1} - \mathbf{U}^{k}\right\|^2_{0,\Omega^S}$$
$$= \Delta t \left\|\mathbf{f}^{S,k+1}\right\|^2_{0,\Omega^S} + \frac{\Delta t}{4} \left\|\frac{\mathbf{U}^{k+1} - \mathbf{U}^{k}}{\Delta t}\right\|^2_{0,\Omega^S}. \quad [6.46]$$

For the fluid, under hypothesis [6.39], we have

$$\left(\mathbf{f}^{F,n+1}, \mathbf{v}^{n+1}\right)_{0,\Omega^F_{n+1}} \leq \frac{1}{\alpha^F} \left\|\mathbf{f}^{F,n+1}\right\|^2_{0,\Omega^F_{n+1}} + \frac{\alpha^F}{4} \left\|\mathbf{v}^{n+1}\right\|^2_{0,\Omega^F_{n+1}}$$
$$\leq \frac{1}{\alpha^F} \left\|\mathbf{f}^{F,n+1}\right\|^2_{0,\Omega^F_{n+1}} + \frac{\alpha^F}{4} \left\|\mathbf{v}^{n+1}\right\|^2_{1,\Omega^F_{n+1}}$$
$$\leq \frac{1}{\alpha^F} \left\|\mathbf{f}^{F,n+1}\right\|^2_{0,\Omega^F_{n+1}} + \frac{1}{4} a^F_{n+1}\left(\mathbf{v}^{n+1}, \mathbf{v}^{n+1}\right) \quad [6.47]$$

and under hypotheses [6.39] and [6.40], we have

$$\left(\mathbf{h}^{F,n+1}, \mathbf{v}^{n+1}\right)_{0,\Gamma^F_N} \leq \frac{C^2_{tr}}{\alpha^F} \left\|\mathbf{h}^{F,n+1}\right\|^2_{0,\Gamma^F_N} + \frac{\alpha^F}{4C^2_{tr}} \left\|\mathbf{v}^{n+1}\right\|^2_{0,\Gamma^F_N}$$
$$\leq \frac{C^2_{tr}}{\alpha^F} \left\|\mathbf{h}^{F,n+1}\right\|^2_{0,\Gamma^F_N} + \frac{\alpha^F}{4} \left\|\mathbf{v}^{n+1}\right\|^2_{1,\Omega^F_{n+1}}$$
$$\leq \frac{C^2_{tr}}{\alpha^F} \left\|\mathbf{h}^{F,n+1}\right\|^2_{0,\Gamma^F_N} + \frac{1}{4} a^F_{n+1}\left(\mathbf{v}^{n+1}, \mathbf{v}^{n+1}\right). \quad [6.48]$$

In [6.43], we use [6.45], [6.47], [6.48], and we obtain

$$\frac{\rho^S}{2}\left\|\frac{\mathbf{U}^{n+1}-\mathbf{U}^n}{\Delta t}\right\|_{0,\Omega^S}^2-\frac{\rho^S}{2}\left\|\frac{\mathbf{U}^{n}-\mathbf{U}^{n-1}}{\Delta t}\right\|_{0,\Omega^S}^2+\frac{1}{2}a^S\left(\mathbf{U}^{n+1},\mathbf{U}^{n+1}\right)$$
$$-\frac{1}{2}a^S\left(\mathbf{U}^{n},\mathbf{U}^{n}\right)+\frac{1}{2}a^S\left(\mathbf{U}^{n+1}-\mathbf{U}^n,\mathbf{U}^{n+1}-\mathbf{U}^n\right)$$
$$+\frac{\rho^F}{2}\int_{\Omega^F_{n+1}}|\mathbf{v}^{n+1}|^2d\mathbf{x}-\frac{\rho^F}{2}\int_{\Omega^F_{n}}|\mathbf{v}^{n}|^2d\mathbf{x}+(\Delta t)a^F_{n+1}(\mathbf{v}^{n+1},\mathbf{v}^{n+1})$$
$$\leq\frac{1}{2\alpha^S}\left\|\mathbf{f}^{S,n+1}\right\|_{0,\Omega^S}^2+\frac{1}{2}a^S\left(\mathbf{U}^{n+1}-\mathbf{U}^n,\mathbf{U}^{n+1}-\mathbf{U}^n\right)$$
$$+\frac{\Delta t}{\alpha^F}\left\|\mathbf{f}^{F,n+1}\right\|_{0,\Omega^F_{n+1}}^2+\frac{(\Delta t)C_{tr}^2}{\alpha^F}\left\|\mathbf{h}^{F,n+1}\right\|_{0,\Gamma^F_N}^2$$
$$+\frac{\Delta t}{2}a^F_{n+1}\left(\mathbf{v}^{n+1},\mathbf{v}^{n+1}\right).\qquad [6.49]$$

In [6.43], we replace n with k and using [6.46], [6.47], [6.48], we obtain

$$\frac{\rho^S}{2}\left\|\frac{\mathbf{U}^{k+1}-\mathbf{U}^k}{\Delta t}\right\|_{0,\Omega^S}^2-\frac{\rho^S}{2}\left\|\frac{\mathbf{U}^{k}-\mathbf{U}^{k-1}}{\Delta t}\right\|_{0,\Omega^S}^2+\frac{1}{2}a^S\left(\mathbf{U}^{k+1},\mathbf{U}^{k+1}\right)$$
$$-\frac{1}{2}a^S\left(\mathbf{U}^{k},\mathbf{U}^{k}\right)+\frac{\rho^F}{2}\int_{\Omega^F_{k+1}}|\mathbf{v}^{k+1}|^2d\mathbf{x}-\frac{\rho^F}{2}\int_{\Omega^F_{k}}|\mathbf{v}^{k}|^2d\mathbf{x}$$
$$+(\Delta t)a^F_{k+1}(\mathbf{v}^{k+1},\mathbf{v}^{k+1})\leq\Delta t\left\|\mathbf{f}^{S,k+1}\right\|_{0,\Omega^S}^2+\frac{\Delta t}{4}\left\|\frac{\mathbf{U}^{k+1}-\mathbf{U}^k}{\Delta t}\right\|_{0,\Omega^S}^2$$
$$+\frac{\Delta t}{\alpha^F}\left\|\mathbf{f}^{F,k+1}\right\|_{0,\Omega^F_{k+1}}^2+\frac{(\Delta t)C_{tr}^2}{\alpha^F}\left\|\mathbf{h}^{F,k+1}\right\|_{0,\Gamma^F_N}^2$$
$$+\frac{\Delta t}{2}a^F_{k+1}\left(\mathbf{v}^{k+1},\mathbf{v}^{k+1}\right).\qquad [6.50]$$

By summing [6.49] with [6.50] for $0\leq k\leq n-1$, we have

$$\frac{\rho^S}{2}\left\|\frac{\mathbf{U}^{n+1}-\mathbf{U}^n}{\Delta t}\right\|_{0,\Omega^S}^2-\frac{\rho^S}{2}\left\|\frac{\mathbf{U}^{0}-\mathbf{U}^{-1}}{\Delta t}\right\|_{0,\Omega^S}^2+\frac{1}{2}a^S\left(\mathbf{U}^{n+1},\mathbf{U}^{n+1}\right)$$
$$-\frac{1}{2}a^S\left(\mathbf{U}^{0},\mathbf{U}^{0}\right)+\frac{\rho^F}{2}\int_{\Omega^F_{n+1}}|\mathbf{v}^{n+1}|^2d\mathbf{x}-\frac{\rho^F}{2}\int_{\Omega^F_{0}}|\mathbf{v}^{0}|^2d\mathbf{x}$$

$$+\frac{\Delta t}{2}\sum_{k=0}^{n} a_{k+1}^{F}(\mathbf{v}^{k+1},\mathbf{v}^{k+1}) \leq \frac{1}{2\alpha^{S}}\left\|\mathbf{f}^{S,n+1}\right\|_{0,\Omega^{S}}^{2}$$
$$+\Delta t\sum_{k=0}^{n-1}\left\|\mathbf{f}^{S,k+1}\right\|_{0,\Omega^{S}}^{2}+\frac{\Delta t}{4}\sum_{k=0}^{n-1}\left\|\frac{\mathbf{U}^{k+1}-\mathbf{U}^{k}}{\Delta t}\right\|_{0,\Omega^{S}}^{2}$$
$$+\frac{\Delta t}{\alpha^{F}}\sum_{k=0}^{n}\left\|\mathbf{f}^{F,k+1}\right\|_{0,\Omega_{k+1}^{F}}^{2}+\frac{(\Delta t)C_{tr}^{2}}{\alpha^{F}}\sum_{k=0}^{n}\left\|\mathbf{h}^{F,k+1}\right\|_{0,\Gamma_{N}^{F}}^{2}. \qquad [6.51]$$

We note

$$\phi_{n+1}=\frac{\rho^{S}}{2}\left\|\frac{\mathbf{U}^{n+1}-\mathbf{U}^{n}}{\Delta t}\right\|_{0,\Omega^{S}}^{2}+\frac{1}{2}a^{S}\left(\mathbf{U}^{n+1},\mathbf{U}^{n+1}\right)+\frac{\rho^{F}}{2}\int_{\Omega_{n+1}^{F}}|\mathbf{v}^{n+1}|^{2}d\mathbf{x}$$
$$+\frac{\Delta t}{2}\sum_{k=0}^{n}a_{k+1}^{F}(\mathbf{v}^{k+1},\mathbf{v}^{k+1})+\frac{\Delta t}{2}a_{0}^{F}(\mathbf{v}^{0},\mathbf{v}^{0}),\quad n\geq 0,$$
$$\phi_{0}=\frac{\rho^{S}}{2}\left\|\frac{\mathbf{U}^{0}-\mathbf{U}^{-1}}{\Delta t}\right\|_{0,\Omega^{S}}^{2}+\frac{1}{2}a^{S}\left(\mathbf{U}^{0},\mathbf{U}^{0}\right)+\frac{\rho^{F}}{2}\int_{\Omega_{0}^{F}}|\mathbf{v}^{0}|^{2}d\mathbf{x}$$
$$+\frac{\Delta t}{2}a_{0}^{F}(\mathbf{v}^{0},\mathbf{v}^{0})$$

and we observe that $\phi_{n+1}\geq\frac{\rho^{S}}{2}\left\|\frac{\mathbf{U}^{n+1}-\mathbf{U}^{n}}{\Delta t}\right\|_{0,\Omega^{S}}^{2}\geq 0,\, n\in\{-1\}\cup\mathbb{N}$.

We have that

$$\Delta t\sum_{k=0}^{n-1}\left\|\mathbf{f}^{S,k+1}\right\|_{0,\Omega^{S}}^{2}\leq n\Delta t\max_{t\in[0,T]}\left\|\mathbf{f}^{S}(t)\right\|_{0,\Omega^{S}}^{2}\leq T\max_{t\in[0,T]}\left\|\mathbf{f}^{S}(t)\right\|_{0,\Omega^{S}}^{2}$$

and similar for $\mathbf{h}^{F}$.

The method for evaluating the terms concerning $\mathbf{f}^{F}$ is different:

$$\Delta t\sum_{k=0}^{n}\left\|\mathbf{f}^{F,k+1}\right\|_{0,\Omega_{k+1}^{F}}^{2}\leq(n+1)\Delta t|\mathbf{f}^{F}|^{2}\max_{0\leq k\leq n}\int_{\Omega_{k+1}^{F}}1\,d\mathbf{x}$$
$$\leq T|\mathbf{f}^{F}|^{2}\max_{0\leq k\leq n}aire(\Omega_{k+1}^{F})\leq T|\mathbf{f}^{F}|^{2}A_{0}.$$

By noting

$$g_0 = \left(\frac{1}{2\alpha^S} + T\right) \max_{t\in[0,T]} \left\|\mathbf{f}^S(t)\right\|^2_{0,\Omega^S} + \frac{C^2_{tr}T}{\alpha^F} \max_{t\in[0,T]} \left\|\mathbf{h}^F(t)\right\|^2_{0,\Gamma^F_N}$$
$$+\frac{T}{\alpha^F}|\mathbf{f}^F|^2 A_0 + \phi_0,$$

we obtain from [6.51]

$$\phi_{n+1} - \phi_0 \leq g_0 - \phi_0 + \frac{\Delta t}{4}\sum_{k=0}^{n-1}\left\|\frac{\mathbf{U}^{k+1} - \mathbf{U}^k}{\Delta t}\right\|^2_{0,\Omega^S}$$
$$= g_0 - \phi_0 + \frac{\Delta t}{2\rho^S}\sum_{k=0}^{n-1}\frac{\rho^S}{2}\left\|\frac{\mathbf{U}^{k+1} - \mathbf{U}^k}{\Delta t}\right\|^2_{0,\Omega^S}$$
$$\leq g_0 - \phi_0 + \frac{\Delta t}{2\rho^S}\sum_{k=0}^{n-1}\phi_{k+1} \leq g_0 - \phi_0 + \frac{\Delta t}{2\rho^S}\sum_{k=0}^{n}\phi_k$$

and, since $\phi_0 \leq g_0$, we can apply the discrete Gronwall lemma and we have

$$\phi_{n+1} \leq g_0 e^{\frac{T}{2\rho^S}} = C$$

given the anticipated conclusion. Taking account of the initial conditions, we have that

$$\phi_0 \leq \frac{\rho^S}{2}\|\mathbf{U}_1\|^2_{0,\Omega^S} + \frac{1}{2}a^S(\mathbf{U}_0, \mathbf{U}_0) + \frac{\rho^F}{2}\int_{\Omega^F_0}|\mathbf{v}_0|^2 d\mathbf{x} + \frac{T}{2}a^F_0(\mathbf{v}_0, \mathbf{v}_0)$$

thus the constant C is independent of Δt. □

6.3. Discretization in time of the non-conservative form: explicit domain calculation

In this section, we shall use the formalism presented in [SY 08]. We keep the notation $\widehat{\Gamma} =]BC[$ for the reference interface (see the left-hand side of Figure 6.1). In contrast, the reference ALE domain for the fluid $\widehat{\Omega}^F$ will change at each step in time.

We assume that the following are known at time t_n:

– the displacement of the structure $\mathbf{U}^n, \mathbf{U}^{n-1} \in (H^1(\Omega^S))^2$, $\mathbf{U}^n = 0$ on Γ_D^S and $\mathbf{U}^{n-1} = 0$ on Γ_D^S;

– the domain $\widehat{\Omega}^F = \Omega_n^F$ with the boundary $\partial\Omega_n = \overline{\Gamma}_D^F \cup \overline{\Gamma}_N^F \cup \overline{\Gamma}_n$;

– the fluid velocity $\mathbf{v}^n \in (H^1(\Omega_n^F))^2$, such that $\nabla_{\widehat{\mathbf{x}}} \cdot \mathbf{v}^n = 0$ in Ω_n^F, $\mathbf{v}^n = 0$ on Γ_D^F; the initial velocity verifies $\nabla \cdot \mathbf{v}_0 = 0$ in Ω_0^F and $\mathbf{v}_0 = 0$ on Γ_D^F;

– the domain velocity $\boldsymbol{\vartheta}^n \in \left(H^1(\Omega_n^F)\right)^2 \cap \left(C^0(\overline{\Omega}_n^F)\right)^2$, such that $\boldsymbol{\vartheta}^n = \mathbf{v}^n$ on Γ_n, $\boldsymbol{\vartheta}^n = 0$ on $\Gamma_D^F \cup \Gamma_N^F$.

We shall construct $\mathcal{A}_{n+1} : \Omega_n^F \to \mathbb{R}^2$ by

$$\mathcal{A}_{n+1}(\widehat{\mathbf{x}}) = \widehat{\mathbf{x}} + \Delta t \boldsymbol{\vartheta}^n(\widehat{\mathbf{x}}).$$

We note $\Omega_{n+1}^F = \mathcal{A}_{n+1}(\Omega_n^F)$, $\Gamma_{n+1} = \mathcal{A}_{n+1}(\Gamma_n)$ and we have $\mathcal{A}_{n+1}(\widehat{\mathbf{x}}) = \widehat{\mathbf{x}}$ on $\Gamma_D^F \cup \Gamma_N^F$.

The Jacobian is

$$\widehat{J}_{n+1}(\hat{\mathbf{x}}) = \det(\nabla_{\hat{\mathbf{x}}}\mathcal{A}_{n+1}(\hat{\mathbf{x}})) = 1 + \Delta t \nabla_{\hat{\mathbf{x}}} \cdot \boldsymbol{\vartheta}^n(\hat{\mathbf{x}}) + (\Delta t)^2 \det(\nabla_{\hat{\mathbf{x}}}\boldsymbol{\vartheta}^n(\hat{\mathbf{x}})). \qquad [6.52]$$

We are looking to find, at time t_{n+1}: the displacement of the structure $\mathbf{U}^{n+1} \in (H^1(\Omega^S))^2$, the fluid velocity $\mathbf{v}^{n+1} \in (H^1(\Omega_{n+1}^F))^2$ and the fluid pressure $p^{n+1} \in L^2(\Omega_{n+1}^F)$ satisfying

$$\begin{aligned}
&\rho^S \int_{\Omega^S} \frac{\mathbf{U}^{n+1} - 2\mathbf{U}^n + \mathbf{U}^{n-1}}{(\Delta t)^2} \cdot \mathbf{w}^S \, d\mathbf{X} + a^S(\mathbf{U}^{n+1}, \mathbf{w}^S) \\
&+\rho^F \int_{\Omega_n^F} \frac{\widehat{\mathbf{v}}^{n+1} - \mathbf{v}^n}{\Delta t} \cdot \widehat{\mathbf{w}}^F \, d\widehat{\mathbf{x}} + \rho \int_{\Omega_n^F} [((\mathbf{v}^n - \boldsymbol{\vartheta}^n) \cdot \nabla_{\widehat{\mathbf{x}}})\widehat{\mathbf{v}}^{n+1}] \cdot \widehat{\mathbf{w}}^F \, d\widehat{\mathbf{x}} \\
&+\rho^F \frac{\Delta t}{2} \int_{\Omega_n^F} \det(\nabla_{\widehat{\mathbf{x}}}\boldsymbol{\vartheta}^n)\widehat{\mathbf{v}}^{n+1} \cdot \widehat{\mathbf{w}}^F \, d\widehat{\mathbf{x}} + a_{n+1}^F(\mathbf{v}^{n+1}, \mathbf{w}^F) + b_{n+1}(\mathbf{w}^F, p^{n+1}) \\
&= \int_{\Omega^S} \mathbf{f}^{S,n+1} \cdot \mathbf{w}^S \, d\mathbf{X} + \int_{\Omega_{n+1}^F} \mathbf{f}^{F,n+1} \cdot \mathbf{w}^F \, d\mathbf{x} + \int_{\Gamma_N^F} \mathbf{h}^{F,n+1} \cdot \mathbf{w}^F \, ds,
\end{aligned}$$

$$\forall \mathbf{w}^S \in (H^1(\Omega^S))^2,\ \mathbf{w}^S = 0 \text{ on } \Gamma_D^S,\ \forall \widehat{\mathbf{w}}^F \in (H^1(\Omega_n^F))^2,\ \widehat{\mathbf{w}}^F = 0 \text{ on } \Gamma_D^F,$$

$$\mathbf{w}^S(\mathbf{X}) = \widehat{\mathbf{w}}^F(\mathcal{A}_n \circ \cdots \circ \mathcal{A}_1(\mathbf{X} + \mathbf{U}_0(\mathbf{X}))), \quad \forall \mathbf{X} \in \widehat{\Gamma} \qquad [6.53]$$

$$b_{n+1}(\mathbf{v}^{n+1}, q) = 0, \quad \forall q \in L^2(\Omega_{n+1}^F) \qquad [6.54]$$

the boundary conditions

$$\mathbf{U}^{n+1} = 0, \text{ on } \Gamma_D^S \qquad [6.55]$$

$$\mathbf{v}^{n+1} = 0, \text{ on } \Gamma_D^F \qquad [6.56]$$

$$\widehat{\mathbf{v}}^{n+1}(\mathcal{A}_n \circ \cdots \circ \mathcal{A}_1(\mathbf{X} + \mathbf{U}_0(\mathbf{X}))) = \frac{\mathbf{U}^{n+1}(\mathbf{X}) - \mathbf{U}^n(\mathbf{X})}{\Delta t}, \quad \forall \mathbf{X} \in \widehat{\Gamma} \qquad [6.57]$$

the initial conditions

$$\mathbf{U}^{-1} = \mathbf{U}_0 - \Delta t \mathbf{U}_1, \qquad [6.58]$$

$$\mathbf{U}^0 = \mathbf{U}_0, \qquad [6.59]$$

$$\mathbf{v}^0 = \mathbf{v}_0, \qquad [6.60]$$

$$\Delta \boldsymbol{\vartheta}^0 = 0, \text{ in } \Omega_0^F,\ \boldsymbol{\vartheta}^0 = \mathbf{v}_0 \text{ on } \Gamma_0 = (\mathbf{id} + \mathbf{U}_0)(\widehat{\Gamma}), \text{ and}$$

$$\boldsymbol{\vartheta}^0 = 0 \text{ on } \Gamma_D^F \cup \Gamma_N^F, \qquad [6.61]$$

where

$$a^S(\mathbf{U}^{n+1}, \mathbf{w}^S) = \lambda^S \int_{\Omega^S} (\nabla_{\mathbf{X}} \cdot \mathbf{U}^{n+1})(\nabla_{\mathbf{X}} \cdot \mathbf{w}^S)\, d\mathbf{X} + 2\mu^S \int_{\Omega^S} \epsilon_{\mathbf{X}}\left(\mathbf{U}^{n+1}\right) : \epsilon_{\mathbf{X}}\left(\mathbf{w}^S\right) d\mathbf{X},$$

$$a_{n+1}^F(\mathbf{v}^{n+1}, \mathbf{w}^F) = \int_{\Omega_{n+1}^F} 2\mu^F\, \boldsymbol{\epsilon}\left(\mathbf{v}^{n+1}\right) : \boldsymbol{\epsilon}\left(\mathbf{w}^F\right) d\mathbf{x},$$

$$b_{n+1}(\mathbf{v}^{n+1}, q) = -\int_{\Omega_{n+1}^F} \left(\nabla \cdot \mathbf{v}^{n+1}\right) q\, d\mathbf{x}.$$

We have

$$\mathbf{v}^{n+1} = \widehat{\mathbf{v}}^{n+1} \circ \mathcal{A}_{n+1}^{-1} \in (H^1(\Omega_{n+1}^F))^2 \text{ and } \mathbf{w}^F = \widehat{\mathbf{w}}^F \circ \mathcal{A}_{n+1}^{-1} \in (H^1(\Omega_{n+1}^F))^2.$$

We can construct $\boldsymbol{\vartheta}^{n+1} \in (H^1(\Omega_{n+1}^F))^2$ by harmonic extension

$$\Delta\boldsymbol{\vartheta}^{n+1} = 0 \text{ in } \Omega_{n+1}^F, \; \boldsymbol{\vartheta}^{n+1} = 0 \text{ on } \Gamma_N^F \cup \Gamma_D^F, \; \boldsymbol{\vartheta}^{n+1} = \mathbf{v}^{n+1} \text{ on } \Gamma_{n+1}. \quad [6.62]$$

Now we can again solve scheme [6.53]–[6.57] to obtain the unknowns at time t_{n+2}.

REMARK 6.1.– *We observe that the boundary Γ_{n+1} is calculated explicitly from Ω_n^F and $\boldsymbol{\vartheta}^n$, and no longer depends on $\mathbf{U}^{n+1}$. At each step in time, the linear system [6.53]–[6.54] must be solved with the boundary conditions [6.55]–[6.57] to obtain $\mathbf{U}^{n+1}$, $\mathbf{v}^{n+1}$ and p^{n+1}. We say that the scheme is semi-implicit.*

For the stability result, we note $|\mathbf{v}^{n+1}|^2 = \mathbf{v}^{n+1} \cdot \mathbf{v}^{n+1}$.

THEOREM 6.2.– *We work under hypotheses [6.39], [6.40] and [6.41].*

i) If $\mathbf{f}^S = 0$, $\mathbf{f}^F = 0$, $\mathbf{h}^F = 0$, then

$$\frac{\rho^S}{2}\left\|\frac{\mathbf{U}^{n+1} - \mathbf{U}^n}{\Delta t}\right\|_{0,\Omega^S}^2 + \frac{1}{2}a^S\left(\mathbf{U}^{n+1}, \mathbf{U}^{n+1}\right) + \frac{\rho^F}{2}\int_{\Omega_{n+1}^F} |\mathbf{v}^{n+1}|^2 d\mathbf{x}$$
$$\leq \frac{\rho^S}{2}\|\mathbf{U}_1\|_{0,\Omega^S}^2 + \frac{1}{2}a^S(\mathbf{U}_0, \mathbf{U}_0) + \frac{\rho^F}{2}\int_{\Omega_0^F} |\mathbf{v}_0|^2 d\mathbf{x}. \quad [6.63]$$

ii) For arbitrary $\mathbf{f}^S \in C^0\left([0,T], (L^2(\Omega^S))^2\right)$, $\mathbf{h}^F \in C^0\left([0,T], (L^2(\Gamma_N^F))^2\right)$, $\mathbf{f}^F(\mathbf{x}, t) = const$ and $area(\Omega_{k+1}^F) < A_0$, for $0 \leq k \leq n$, hence there exists a constant $C > 0$, such that for $0 \leq n \leq N-1$, we have

$$\frac{\rho^S}{2}\left\|\frac{\mathbf{U}^{n+1} - \mathbf{U}^n}{\Delta t}\right\|_{0,\Omega^S}^2 + \frac{1}{2}a^S\left(\mathbf{U}^{n+1}, \mathbf{U}^{n+1}\right) + \frac{\rho^F}{2}\int_{\Omega_{n+1}^F} |\mathbf{v}^{n+1}|^2 d\mathbf{x}$$
$$+ \frac{\Delta t}{2}\sum_{k=0}^{n+1} a_k^F(\mathbf{v}^k, \mathbf{v}^k) \leq C.$$

DEMONSTRATION 6.2.– We set $\mathbf{w}^S = \mathbf{U}^{n+1} - \mathbf{U}^n$ and $\widehat{\mathbf{w}}^F = \Delta t \widehat{\mathbf{v}}^{n+1}$. Taking account of [6.55]–[6.57], we have that $\mathbf{w}^S$, $\widehat{\mathbf{w}}^F$ are admissible test functions and by substituting them into [6.53], we obtain

$$\begin{aligned}
&\rho^S \int_{\Omega^S} \frac{\mathbf{U}^{n+1} - 2\mathbf{U}^n + \mathbf{U}^{n-1}}{(\Delta t)^2} \cdot (\mathbf{U}^{n+1} - \mathbf{U}^n) d\mathbf{X} + a^S(\mathbf{U}^{n+1}, \mathbf{U}^{n+1} - \mathbf{U}^n) \\
&+ \rho^F \int_{\Omega_n^F} (\widehat{\mathbf{v}}^{n+1} - \mathbf{v}^n) \cdot \widehat{\mathbf{v}}^{n+1} d\mathbf{x} \\
&+ \Delta t \rho^F \int_{\Omega_n^F} [((\mathbf{v}^n - \boldsymbol{\vartheta}^n) \cdot \nabla_{\widehat{\mathbf{x}}}) \widehat{\mathbf{v}}^{n+1}] \cdot \widehat{\mathbf{v}}^{n+1} \, d\widehat{\mathbf{x}} \\
&+ \frac{(\Delta t)^2}{2} \rho^F \int_{\Omega_n^F} \det(\nabla_{\widehat{\mathbf{x}}} \boldsymbol{\vartheta}^n) \widehat{\mathbf{v}}^{n+1} \cdot \widehat{\mathbf{v}}^{n+1} \, d\widehat{\mathbf{x}} + (\Delta t) a_{n+1}^F(\mathbf{v}^{n+1}, \mathbf{v}^{n+1}) \\
&+ (\Delta t) b_{n+1}(\mathbf{v}^{n+1}, p^{n+1}) = \int_{\Omega^S} \mathbf{f}^{S,n+1} \cdot (\mathbf{U}^{n+1} - \mathbf{U}^n) \, d\mathbf{X} \\
&+ \Delta t \int_{\Omega_{n+1}^F} \mathbf{f}^{F,n+1} \cdot \mathbf{v}^{n+1} \, d\mathbf{x} + \Delta t \int_{\Gamma_N^F} \mathbf{h}^{F,n+1} \cdot \mathbf{v}^{n+1} ds. \qquad [6.64]
\end{aligned}$$

Apart from the second and third lines of the above equation, all of the terms will be treated as in theorem 6.1.

Using [4.27] and [4.28], we obtain

$$\begin{aligned}
&\rho^S \int_{\Omega^S} \frac{\mathbf{U}^{n+1} - 2\mathbf{U}^n + \mathbf{U}^{n-1}}{(\Delta t)^2} \cdot (\mathbf{U}^{n+1} - \mathbf{U}^n) d\mathbf{X} + a^S(\mathbf{U}^{n+1}, \mathbf{U}^{n+1} - \mathbf{U}^n) \\
&\geq \frac{\rho^S}{2} \left\| \frac{\mathbf{U}^{n+1} - \mathbf{U}^n}{\Delta t} \right\|_{0,\Omega^S}^2 - \frac{\rho^S}{2} \left\| \frac{\mathbf{U}^n - \mathbf{U}^{n-1}}{\Delta t} \right\|_{0,\Omega^S}^2 + \frac{1}{2} a^S \left(\mathbf{U}^{n+1}, \mathbf{U}^{n+1} \right) \\
&- \frac{1}{2} a^S (\mathbf{U}^n, \mathbf{U}^n) + \frac{1}{2} a^S \left(\mathbf{U}^{n+1} - \mathbf{U}^n, \mathbf{U}^{n+1} - \mathbf{U}^n \right).
\end{aligned}$$

As in the demonstration of proposition 3.8, we have

$$\int_{\Omega_n^F} \mathbf{v}^n \cdot \widehat{\mathbf{v}}^{n+1} \, d\mathbf{x} \leq \frac{1}{2} \int_{\Omega_n^F} |\mathbf{v}^n|^2 d\mathbf{x} + \frac{1}{2} \int_{\Omega_n^F} |\widehat{\mathbf{v}}^{n+1}|^2 d\mathbf{x}$$

and by using $[(\mathbf{w}\cdot\nabla)\mathbf{v}]\cdot\mathbf{v} = \frac{1}{2}\mathbf{w}\cdot(\nabla|\mathbf{v}|^2)$, we obtain

$$\int_{\Omega_n^F} [((\mathbf{v}^n - \boldsymbol{\vartheta}^n)\cdot\nabla_{\widehat{\mathbf{x}}})\widehat{\mathbf{v}}^{n+1}]\cdot\widehat{\mathbf{v}}^{n+1}\,d\widehat{\mathbf{x}} = \frac{1}{2}\int_{\Omega_n^F}(\mathbf{v}^n - \boldsymbol{\vartheta}^n)\cdot(\nabla_{\widehat{\mathbf{x}}}|\widehat{\mathbf{v}}^{n+1}|^2)\,d\widehat{\mathbf{x}}$$
$$= \frac{1}{2}\int_{\partial\Omega_n^F}(\mathbf{v}^n - \boldsymbol{\vartheta}^n)\cdot\mathbf{n}^F|\widehat{\mathbf{v}}^{n+1}|^2 ds - \frac{1}{2}\int_{\Omega_n^F}\nabla_{\widehat{\mathbf{x}}}\cdot(\mathbf{v}^n - \boldsymbol{\vartheta}^n)|\widehat{\mathbf{v}}^{n+1}|^2 d\widehat{\mathbf{x}}$$
$$= \frac{1}{2}\int_{\Gamma_N^F}\mathbf{v}^n\cdot\mathbf{n}^F|\widehat{\mathbf{v}}^{n+1}|^2 ds + \frac{1}{2}\int_{\Omega_n^F}(\nabla_{\widehat{\mathbf{x}}}\cdot\boldsymbol{\vartheta}^n)|\widehat{\mathbf{v}}^{n+1}|^2 d\widehat{\mathbf{x}}.$$

For the last equation, we have used the fact that $\nabla_{\widehat{\mathbf{x}}}\cdot\mathbf{v}^n = 0$ in Ω_n^F, and the boundary conditions $\mathbf{v}^n = \boldsymbol{\vartheta}^n$ on Γ_n, $\boldsymbol{\vartheta}^n = 0$ on $\Gamma_N^F \cup \Gamma_D^F$ and $\mathbf{v}^n = 0$ on Γ_D^F.

We follow up with

$$\rho^F\int_{\Omega_n^F}(\widehat{\mathbf{v}}^{n+1} - \mathbf{v}^n)\cdot\widehat{\mathbf{v}}^{n+1} d\mathbf{x}$$
$$+ \Delta t\rho^F\int_{\Omega_n^F}[((\mathbf{v}^n - \boldsymbol{\vartheta}^n)\cdot\nabla_{\widehat{\mathbf{x}}})\widehat{\mathbf{v}}^{n+1}]\cdot\widehat{\mathbf{v}}^{n+1}\,d\widehat{\mathbf{x}}$$
$$+ \frac{(\Delta t)^2}{2}\rho^F\int_{\Omega_n^F}\det(\nabla_{\widehat{\mathbf{x}}}\boldsymbol{\vartheta}^n)\widehat{\mathbf{v}}^{n+1}\cdot\widehat{\mathbf{v}}^{n+1}\,d\widehat{\mathbf{x}}$$
$$\geq \rho^F\int_{\Omega_n^F}|\widehat{\mathbf{v}}^{n+1}|^2 d\widehat{\mathbf{x}} - \frac{\rho^F}{2}\int_{\Omega_n^F}|\mathbf{v}^n|^2 d\mathbf{x} - \frac{\rho^F}{2}\int_{\Omega_n^F}|\widehat{\mathbf{v}}^{n+1}|^2 d\mathbf{x}$$
$$+ \frac{\Delta t\rho^F}{2}\int_{\Gamma_N^F}\mathbf{v}^n\cdot\mathbf{n}^F|\widehat{\mathbf{v}}^{n+1}|^2 ds + \frac{\Delta t\rho^F}{2}\int_{\Omega_n^F}(\nabla_{\widehat{\mathbf{x}}}\cdot\boldsymbol{\vartheta}^n)|\widehat{\mathbf{v}}^{n+1}|^2 d\widehat{\mathbf{x}}$$
$$+ \frac{(\Delta t)^2}{2}\rho^F\int_{\Omega_n^F}\det(\nabla_{\widehat{\mathbf{x}}}\boldsymbol{\vartheta}^n)|\widehat{\mathbf{v}}^{n+1}|^2\,d\widehat{\mathbf{x}}.$$

We shall use [6.52] and the fact that $\int_{\Omega_n^F}|\widehat{\mathbf{v}}^{n+1}|^2\widehat{J}_{n+1}(\widehat{\mathbf{x}})d\widehat{\mathbf{x}} = \int_{\Omega_{n+1}^F}|\mathbf{v}^{n+1}|^2 d\mathbf{x}$ and we obtain

$$\frac{\rho^F}{2}\int_{\Omega_n^F}\left(1 + \Delta t(\nabla_{\widehat{\mathbf{x}}}\cdot\boldsymbol{\vartheta}^n) + (\Delta t)^2\det(\nabla_{\widehat{\mathbf{x}}}\boldsymbol{\vartheta}^n)\right)|\widehat{\mathbf{v}}^{n+1}|^2 d\mathbf{x}$$
$$- \frac{\rho^F}{2}\int_{\Omega_n^F}|\mathbf{v}^n|^2 d\mathbf{x} + \frac{\Delta t\rho^F}{2}\int_{\Gamma_N^F}\mathbf{v}^n\cdot\mathbf{n}^F|\mathbf{v}^{n+1}|^2 ds$$

$$= \frac{\rho^F}{2}\int_{\Omega^F_{n+1}} |\mathbf{v}^{n+1}|^2 d\mathbf{x} - \frac{\rho^F}{2}\int_{\Omega^F_n} |\mathbf{v}^n|^2 d\mathbf{x}$$

$$+ \frac{\Delta t \rho^F}{2}\int_{\Gamma^F_N} \mathbf{v}^n \cdot \mathbf{n} |\mathbf{v}^{n+1}|^2 ds \geq \frac{\rho^F}{2}\int_{\Omega^F_{n+1}} |\mathbf{v}^{n+1}|^2 d\mathbf{x} - \frac{\rho^F}{2}\int_{\Omega^F_n} |\mathbf{v}^n|^2 d\mathbf{x}.$$

Given that $b_{n+1}(\mathbf{v}^{n+1}, p^{n+1}) = 0$, we obtain from [6.64] that

$$\frac{\rho^S}{2}\left\|\frac{\mathbf{U}^{n+1}-\mathbf{U}^n}{\Delta t}\right\|^2_{0,\Omega^S} - \frac{\rho^S}{2}\left\|\frac{\mathbf{U}^{n}-\mathbf{U}^{n-1}}{\Delta t}\right\|^2_{0,\Omega^S} + \frac{1}{2}a^S\left(\mathbf{U}^{n+1}, \mathbf{U}^{n+1}\right)$$

$$-\frac{1}{2}a^S\left(\mathbf{U}^{n}, \mathbf{U}^{n}\right) + \frac{1}{2}a^S\left(\mathbf{U}^{n+1}-\mathbf{U}^n, \mathbf{U}^{n+1}-\mathbf{U}^n\right)$$

$$+\frac{\rho^F}{2}\int_{\Omega^F_{n+1}} |\mathbf{v}^{n+1}|^2 d\mathbf{x} - \frac{\rho^F}{2}\int_{\Omega^F_n} |\mathbf{v}^n|^2 d\mathbf{x} + (\Delta t) a^F_{n+1}(\mathbf{v}^{n+1}, \mathbf{v}^{n+1})$$

$$\leq \int_{\Omega^S} \mathbf{f}^{S,n+1}\cdot(\mathbf{U}^{n+1}-\mathbf{U}^n) d\mathbf{X} + \Delta t\int_{\Omega^F_{n+1}} \mathbf{f}^{F,n+1}\cdot \mathbf{v}^{n+1} d\mathbf{x}$$

$$+\Delta t\int_{\Gamma^F_N} \mathbf{h}^{F,n+1}\cdot \mathbf{v}^{n+1} ds. \qquad [6.65]$$

i) If $\mathbf{f}^S = 0, \mathbf{f}^F = 0, \mathbf{h}^F = 0$, then

$$\frac{\rho^S}{2}\left\|\frac{\mathbf{U}^{n+1}-\mathbf{U}^n}{\Delta t}\right\|^2_{0,\Omega^S} - \frac{\rho^S}{2}\left\|\frac{\mathbf{U}^{n}-\mathbf{U}^{n-1}}{\Delta t}\right\|^2_{0,\Omega^S} + \frac{1}{2}a^S\left(\mathbf{U}^{n+1}, \mathbf{U}^{n+1}\right)$$

$$-\frac{1}{2}a^S\left(\mathbf{U}^{n}, \mathbf{U}^{n}\right) + \frac{\rho^F}{2}\int_{\Omega^F_{n+1}} |\mathbf{v}^{n+1}|^2 d\mathbf{x} - \frac{\rho^F}{2}\int_{\Omega^F_n} |\mathbf{v}^n|^2 d\mathbf{x} \leq 0 \qquad [6.66]$$

thus

$$\frac{\rho^S}{2}\left\|\frac{\mathbf{U}^{n+1}-\mathbf{U}^n}{\Delta t}\right\|^2_{0,\Omega^S} + \frac{1}{2}a^S\left(\mathbf{U}^{n+1}, \mathbf{U}^{n+1}\right) + \frac{\rho^F}{2}\int_{\Omega^F_{n+1}} |\mathbf{v}^{n+1}|^2 d\mathbf{x}$$

$$\leq \frac{\rho^S}{2}\left\|\frac{\mathbf{U}^{0}-\mathbf{U}^{-1}}{\Delta t}\right\|^2_{0,\Omega^S} + \frac{1}{2}a^S\left(\mathbf{U}^{0}, \mathbf{U}^{0}\right) + \frac{\rho^F}{2}\int_{\Omega^F_0} |\mathbf{v}^0|^2 d\mathbf{x}.$$

Taking account of the initial conditions [6.58]–[6.60], we have [6.63].

ii) We shall expand the right-hand side of [6.65]. For the structure, we have

$$\left(\mathbf{f}^{S,n+1}, \mathbf{U}^{n+1} - \mathbf{U}^{n}\right)_{0,\Omega^S} \leq \frac{1}{2\alpha^S} \left\|\mathbf{f}^{S,n+1}\right\|_{0,\Omega^S}^2 + \frac{1}{2} a^S \left(\mathbf{U}^{n+1} - \mathbf{U}^n, \mathbf{U}^{n+1} - \mathbf{U}^n\right)$$

and for $0 \leq k \leq n-1$,

$$\left(\mathbf{f}^{S,k+1}, \mathbf{U}^{k+1} - \mathbf{U}^{k}\right)_{0,\Omega^S} \leq \Delta t \left\|\mathbf{f}^{S,k+1}\right\|_{0,\Omega^S}^2 + \frac{\Delta t}{4} \left\|\frac{\mathbf{U}^{k+1} - \mathbf{U}^k}{\Delta t}\right\|_{0,\Omega^S}^2.$$

For the fluid, under hypotheses [6.39] and [6.40], we have

$$\left(\mathbf{f}^{F,n+1}, \mathbf{v}^{n+1}\right)_{0,\Omega^F_{n+1}} \leq \frac{1}{\alpha^F} \left\|\mathbf{f}^{F,n+1}\right\|_{0,\Omega^F_{n+1}}^2 + \frac{1}{4} a^F_{n+1} \left(\mathbf{v}^{n+1}, \mathbf{v}^{n+1}\right)$$

$$\left(\mathbf{h}^{F,n+1}, \mathbf{v}^{n+1}\right)_{0,\Gamma^F_N} \leq \frac{C_{tr}^2}{\alpha^F} \left\|\mathbf{h}^{F,n+1}\right\|_{0,\Gamma^F_N}^2 + \frac{1}{4} a^F_{n+1} \left(\mathbf{v}^{n+1}, \mathbf{v}^{n+1}\right).$$

We note

$$\phi_{n+1} = \frac{\rho^S}{2} \left\|\frac{\mathbf{U}^{n+1} - \mathbf{U}^n}{\Delta t}\right\|_{0,\Omega^S}^2 + \frac{1}{2} a^S \left(\mathbf{U}^{n+1}, \mathbf{U}^{n+1}\right) + \frac{\rho^F}{2} \int_{\Omega^F_{n+1}} |\mathbf{v}^{n+1}|^2 d\mathbf{x}$$
$$+ \frac{\Delta t}{2} \sum_{k=0}^{n} a^F_{k+1}(\mathbf{v}^{k+1}, \mathbf{v}^{k+1}) + \frac{\Delta t}{2} a^F_0(\mathbf{v}^0, \mathbf{v}^0), \quad n \geq 0,$$

$$\phi_0 = \frac{\rho^S}{2} \left\|\frac{\mathbf{U}^0 - \mathbf{U}^{-1}}{\Delta t}\right\|_{0,\Omega^S}^2 + \frac{1}{2} a^S \left(\mathbf{U}^0, \mathbf{U}^0\right) + \frac{\rho^F}{2} \int_{\Omega^F_0} |\mathbf{v}^0|^2 d\mathbf{x}$$
$$+ \frac{\Delta t}{2} a^F_0(\mathbf{v}^0, \mathbf{v}^0).$$

Taking account of the expansion of the right-hand side of [6.65], we have

$$\phi_{n+1} - \phi_n \leq \frac{1}{2\alpha^S} \left\|\mathbf{f}^{S,n+1}\right\|_{0,\Omega^S}^2 + \frac{\Delta t}{\alpha^F} \left\|\mathbf{f}^{F,n+1}\right\|_{0,\Omega^F_{n+1}}^2$$
$$+ \frac{(\Delta t) C_{tr}^2}{\alpha^F} \left\|\mathbf{h}^{F,n+1}\right\|_{0,\Gamma^F_N}^2$$

and for $0 \leq k \leq n-1$

$$\phi_{k+1} - \phi_k \leq \Delta t \left\| \mathbf{f}^{S,k+1} \right\|_{0,\Omega^S}^2 + \frac{\Delta t}{4} \left\| \frac{\mathbf{U}^{k+1} - \mathbf{U}^k}{\Delta t} \right\|_{0,\Omega^S}^2 + \frac{\Delta t}{\alpha^F} \left\| \mathbf{f}^{F,k+1} \right\|_{0,\Omega_{k+1}^F}^2 + \frac{(\Delta t) C_{tr}^2}{\alpha^F} \left\| \mathbf{h}^{F,k+1} \right\|_{0,\Gamma_N^F}^2 .$$

Summing, we obtain

$$\phi_{n+1} - \phi_0 \leq \frac{1}{2\alpha^S} \left\| \mathbf{f}^{S,n+1} \right\|_{0,\Omega^S}^2 + \Delta t \sum_{k=0}^{n-1} \left\| \mathbf{f}^{S,k+1} \right\|_{0,\Omega^S}^2 + \frac{\Delta t}{4} \sum_{k=0}^{n-1} \left\| \frac{\mathbf{U}^{k+1} - \mathbf{U}^k}{\Delta t} \right\|_{0,\Omega^S}^2 + \frac{\Delta t}{\alpha^F} \sum_{k=0}^{n} \left\| \mathbf{f}^{F,k+1} \right\|_{0,\Omega_{k+1}^F}^2 + \frac{(\Delta t) C_{tr}^2}{\alpha^F} \sum_{k=0}^{n} \left\| \mathbf{h}^{F,k+1} \right\|_{0,\Gamma_N^F}^2 .$$

Noting that

$$g_0 = \left(\frac{1}{2\alpha^S} + T \right) \max_{t \in [0,T]} \left\| \mathbf{f}^S(t) \right\|_{0,\Omega^S}^2 + \frac{C_{tr}^2 T}{\alpha^F} \max_{t \in [0,T]} \left\| \mathbf{h}^F(t) \right\|_{0,\Gamma_N^F}^2 + \frac{T}{\alpha^F} |\mathbf{f}^F|^2 A_0 + \phi_0,$$

we have

$$\phi_{n+1} \leq g_0 + \frac{\Delta t}{4} \sum_{k=0}^{n-1} \left\| \frac{\mathbf{U}^{k+1} - \mathbf{U}^k}{\Delta t} \right\|_{0,\Omega^S}^2 \leq g_0 + \frac{\Delta t}{2\rho^S} \sum_{k=0}^{n} \phi_k$$

and, since $\phi_0 \leq g_0$, we can apply the discrete Gronwall lemma and we have

$$\phi_{n+1} \leq g_0 e^{\frac{T}{2\rho^S}} = C$$

giving the desired conclusion. Taking account of the initial conditions, we have that

$$\phi_0 \leq \frac{\rho^S}{2} \|\mathbf{U}_1\|_{0,\Omega^S}^2 + \frac{1}{2} a^S(\mathbf{U}_0, \mathbf{U}_0) + \frac{\rho^F}{2} \int_{\Omega_0^F} |\mathbf{v}_0|^2 d\mathbf{x} + \frac{T}{2} a_0^F(\mathbf{v}_0, \mathbf{v}_0)$$

thus the constant C is independent of Δt. □

6.4. Coupling strategies

We have seen that using the implicit Euler scheme for the fluid and the backward Euler scheme for the structure, we obtain a scheme for solving the fluid–structure interaction problem that is stable independently of Δt. When we studied the stability of the implicit Euler scheme for the fluid, we used $\mathbf{w}^F = \Delta t \mathbf{v}^{n+1}$ as the test function. For the backward Euler scheme for the structure, we used $\mathbf{w}^S = \mathbf{U}^{n+1} - \mathbf{U}^n$. The key to obtaining a stable scheme was to choose the condition

$$\mathbf{v}^{n+1} = \frac{\mathbf{U}^{n+1} - \mathbf{U}^n}{\Delta t}$$

at the interface, and in this case $\mathbf{w}^S, \mathbf{w}^F$ are admissible test functions for the fluid–structure interaction problem.

Using the same strategy, we can construct other stable schemes for the fluid–structure interaction:

– the implicit Euler scheme for the fluid and the implicit centered scheme for the structure with the interface condition

$$\mathbf{v}^{n+1} = \frac{\mathbf{U}^{n+1} - \mathbf{U}^{n-1}}{2\Delta t};$$

– the implicit Euler scheme for the fluid and the mid-point scheme for the structure with the interface condition

$$\mathbf{v}^{n+1} = \frac{\mathbf{U}^{n+1} - \mathbf{U}^n}{\Delta t}.$$

We can also couple an implicit scheme for the fluid with an implicit scheme for the nonlinear elasticity, as in [FER 09].

6.4.1. *Fixed-point algorithm with relaxation: implicit domain calculation*

We know $\mathbf{U}^n$, $\mathbf{U}^{n-1}$, Ω_n^F, $\mathbf{v}^n$, p^n, $\boldsymbol{\vartheta}^n$. The ALE reference domain for the fluid is $\widehat{\Omega}^F$.

We want to find the solution at $t_{n+1} = (n+1)\Delta t$.

The relaxation factor is $0 < \omega \leq 1$ and, for the stopping test, we use $tol > 0$.

– *Step 1.* Initialization $k = 0$, extrapolation of the displacement

$$\mathbf{U}_k^{n+1} = 2\mathbf{U}^n - \mathbf{U}^{n-1}$$

– *Step 2.* Construction of the ALE map, $\mathcal{A}_{n+1,k+1}(\widehat{\mathbf{x}}) = \widehat{\mathbf{x}} + \widehat{\mathbf{d}}_{k+1}^{n+1}(\widehat{\mathbf{x}})$, where

$$\begin{aligned}
\Delta\widehat{\mathbf{d}}_{k+1}^{n+1} &= 0, \text{ in } \widehat{\Omega}^F, \\
\widehat{\mathbf{d}}_{k+1}^{n+1} &= \mathbf{U}_k^{n+1}, \text{ on } \widehat{\Gamma}, \\
\widehat{\mathbf{d}}_{k+1}^{n+1} &= 0, \text{ on } \Gamma_D^F \cup \Gamma_N^F.
\end{aligned}$$

We set

$$\mathcal{A}_{k+1/2}^{n+1} = \frac{1}{2}\mathcal{A}_n + \frac{1}{2}\mathcal{A}_{n+1,k+1}, \quad \widehat{\boldsymbol{\vartheta}}_{k+1}^{n+1} = \frac{\widehat{\mathbf{d}}_{k+1}^{n+1} - \widehat{\mathbf{d}}^n}{\Delta t} = \widehat{\boldsymbol{\vartheta}}_{k+1/2}^{n+1}$$

– *Step 3.* Solve the fluid in $\boldsymbol{\Omega}^{\mathbf{F}}_{\mathbf{n+1,k+1}} = \mathcal{A}_{\mathbf{n+1,k+1}}(\widehat{\boldsymbol{\Omega}}^{\mathbf{F}})$ with the Dirichlet interface condition

$$\mathbf{v}_{k+1}^{n+1}(\mathbf{id} + \mathbf{U}_{k+1}^{n+1}) = \frac{\mathbf{U}_k^{n+1} - \mathbf{U}^n}{\Delta t} \text{ on } \widehat{\Gamma} \Rightarrow \mathbf{v}_{k+1}^{n+1}, p_{k+1}^{n+1}$$

– *Step 4.* Solve the structure with the Neumann interface condition

$$\sigma^F(\mathbf{v}_{k+1}^{n+1}, p_{k+1}^{n+1})|_{\Gamma_{n+1,k+1}} \Rightarrow \widetilde{\mathbf{U}}_{k+1}^{n+1}$$

– *Step 5.* Stopping test

$$\left\|\widetilde{\mathbf{U}}_{k+1}^{n+1} - \mathbf{U}_k^{n+1}\right\|_{0,\Omega^S} < tol \quad \text{or} \quad \frac{\left\|\widetilde{\mathbf{U}}_{k+1}^{n+1} - \mathbf{U}_k^{n+1}\right\|_{0,\Omega^S}}{\left\|\widetilde{\mathbf{U}}_{k+1}^{n+1}\right\|_{0,\Omega^S}} < tol$$

- If the stopping condition does not hold, use the relaxation

$$\mathbf{U}_{k+1}^{n+1} = \omega\widetilde{\mathbf{U}}_{k+1}^{n+1} + (1-\omega)\mathbf{U}_k^{n+1}$$

$k = k + 1$ and go to *Step 2*

- If the stopping condition holds, we set $\mathbf{U}^{n+1} = \mathbf{U}_k^{n+1}$, $\Omega_{n+1}^F = \Omega_{n+1,k+1}^F$, $\mathbf{v}^{n+1} = \mathbf{v}_{k+1}^{n+1}$, $p^{n+1} = p_{k+1}^{n+1}$, $\boldsymbol{\vartheta}^{n+1} = \boldsymbol{\vartheta}_{k+1}^{n+1}$.

This type of algorithm, that uses implicit calculation of the domain, has been used in [NOB 01] and [QUA 04]. A study for the choice of ω is presented in [CAU 05].

6.4.2. *Fixed-point algorithm with relaxation: explicit domain calculation*

We know $\mathbf{U}^n$, $\mathbf{U}^{n-1}$, Ω_n^F, $\mathbf{v}^n$, p^n, $\boldsymbol{\vartheta}^n$.

We want to find the solution at $t_{n+1} = (n+1)\Delta t$.

The relaxation factor is $0 < \omega \leq 1$ and, for the stopping test, we use $tol > 0$.

– *Step 1.* Explicit calculation of the fluid domain. We set $\widehat{\Omega}^F = \Omega_n^F$

$$\mathcal{A}_{n+1}(\widehat{x}) = \widehat{x} + \Delta t \boldsymbol{\vartheta}^n(\widehat{x}),\ \widehat{x} \in \Omega_n^F,\ \mathcal{A}_{n+1}(\Omega_n^F) = \Omega_{n+1}^F,\ \mathcal{A}_{n+1}(\Gamma_n) = \Gamma_{n+1}.$$

– *Step 2.* Initialization $k = 0$, $\mathbf{U}_k^{n+1} = 2\mathbf{U}^n - \mathbf{U}^{n-1}$.

– *Step 3.* Solve the fluid in $\boldsymbol{\Omega}_\mathbf{n}^\mathbf{F}$ with the Dirichlet interface condition

$$\widehat{\mathbf{v}}_{k+1}^{n+1}(\mathcal{A}_n \circ \cdots \circ \mathcal{A}_1(\mathbf{id} + \mathbf{U}_0)) = \frac{\mathbf{U}_k^{n+1} - \mathbf{U}^n}{\Delta t} \text{ on } \widehat{\Gamma} \Rightarrow \widehat{\mathbf{v}}_{k+1}^{n+1}, \widehat{p}_{k+1}^{n+1}$$

– *Step 4.* Solve the structure with the Neumann interface condition

$$\sigma^F(\widehat{\mathbf{v}}_{k+1}^{n+1}, \widehat{p}_{k+1}^{n+1})|_{\Gamma_n} \Rightarrow \widetilde{\mathbf{U}}_{k+1}^{n+1}$$

– *Step 5.* Stopping test

$$\left\| \widetilde{\mathbf{U}}_{k+1}^{n+1} - \mathbf{U}_k^{n+1} \right\|_{0,\Omega^S} < tol \quad \text{or} \quad \frac{\left\| \widetilde{\mathbf{U}}_{k+1}^{n+1} - \mathbf{U}_k^{n+1} \right\|_{0,\Omega^S}}{\left\| \widetilde{\mathbf{U}}_{k+1}^{n+1} \right\|_{0,\Omega^S}} < tol$$

- If the stopping condition does not hold, use the relaxation

$$\mathbf{U}_{k+1}^{n+1} = \omega \widetilde{\mathbf{U}}_{k+1}^{n+1} + (1 - \omega)\mathbf{U}_k^{n+1}$$

$k = k + 1$ and go to *Step 3*

- If the stopping condition holds, we set $\mathbf{U}^{n+1} = \mathbf{U}_{k+1}^{n+1}$, $\mathbf{v}^{n+1} = \widehat{\mathbf{v}}_{k+1}^{n+1}(\mathcal{A}_{n+1}^{-1})$, $p^{n+1} = \widehat{p}_{k+1}^{n+1}(\mathcal{A}_{n+1}^{-1})$ and

$$\begin{aligned} \Delta \boldsymbol{\vartheta}^{n+1} &= 0 \text{ in } \Omega_{n+1}^F, \\ \boldsymbol{\vartheta}^{n+1} &= 0 \text{ on } \Gamma_D^F \cup \Gamma_N^F, \\ \boldsymbol{\vartheta}^{n+1} &= \mathbf{v}^{n+1} \text{ on } \Gamma_{n+1}. \end{aligned}$$

REMARK 6.2.– *When we compute the domain implicitly, we solve the fluid problem in a domain $\Omega_{n+1,k+1}^F$ that is modified at each iteration of k. In contrast, when the domain calculation is explicit, the fluid problem is solved in the domain Ω_n^F, which is independent of k. We reduce the calculation time considerably by adopting an explicit domain calculation strategy.*

REMARK 6.3.– *In the above algorithms, the structure problem is solved separately from the fluid problem so we call them* partitioned procedures. *We can use pre-existing black box computation codes to solve the structure and the fluid, respectively.*

6.5. The constants in the Poincaré, Korn and trace inequalities

To obtain the results for stability in the previous sections, we worked under the hypotheses that the ellipticity constant and the constant in the continuity of the function trace did not depend on the domain motion, [6.39] and [6.40]. In this section, we shall show that the constants in the Poincaré, Korn and trace inequalities are, under certain conditions, independent of the domain.

We shall work with a domain as shown in Figure 3.1. Let L, H be two positive constants. Let $u : [0, L] \to \mathbb{R}$ be a function $u \in C^1([0, L])$, such that $u(0) = u(L) = 0$. Furthermore, we assume that $H + \min_{x_1 \in [0,L]} u(x_1) \geq d_0$, where $d_0 > 0$ is a constant. We shall introduce the domain

$$\Omega_u = \left\{(x_1, x_2) \in \mathbb{R}^2;\ x_1 \in]0, L[,\ 0 < x_2 < H + u(x_1)\right\}$$

which has the boundaries

$$\Gamma_{in} = \left\{(0, x_2) \in \mathbb{R}^2;\ x_2 \in]0, H[\right\}$$
$$\Gamma_D = \left\{(x_1, 0) \in \mathbb{R}^2;\ x_1 \in]0, L[\right\}$$
$$\Gamma_{out} = \left\{(L, x_2) \in \mathbb{R}^2;\ x_2 \in]0, H[\right\}$$
$$\Gamma_u = \left\{(x_1, x_2) \in \mathbb{R}^2;\ x_1 \in]0, L[,\ x_2 = H + u(x_1)\right\}.$$

As $u'(0)$ and $u'(L)$ are neither $+\infty$ nor $-\infty$, the domain Ω_u is Lipschitz.

PROPOSITION 6.1.– *[Poincaré's inequality] If there exists* $d > 0$ *such that* $H + \max_{x_1 \in [0,L]} u(x_1) \leq d$, *then*

$$\forall v \in H^1(\Omega_u),\ v = 0 \ on\ \Gamma_u, \quad \|v\|_{0,\Omega_u} \leq \frac{d}{\sqrt{2}} \|\nabla v\|_{0,\Omega_u}. \qquad [6.67]$$

DEMONSTRATION 6.3.– We shall first demonstrate the Poincaré's inequality for $v \in C^1(\overline{\Omega}_u)$ such that $v = 0$ on Γ_u. As $v(x_1, 0) = 0$, we have the identity

$$v(x_1, x_2) = \int_0^{x_2} \frac{\partial v}{\partial x_2}(x_1, \xi) d\xi$$

and by applying the Cauchy–Schwarz inequality, we obtain

$$|v(x_1, x_2)| \leq \int_0^{x_2} \left|\frac{\partial v}{\partial x_2}(x_1, \xi)\right| d\xi \leq \sqrt{\int_0^{x_2} 1^2 d\xi} \sqrt{\int_0^{x_2} \left|\frac{\partial v}{\partial x_2}(x_1, \xi)\right|^2 d\xi}$$
$$\leq \sqrt{x_2} \sqrt{\int_0^{H+u(x_1)} \left|\frac{\partial v}{\partial x_2}(x_1, \xi)\right|^2 d\xi}$$

thus

$$|v(x_1, x_2)|^2 \leq x_2 \int_0^{H+u(x_1)} \left|\frac{\partial v}{\partial x_2}(x_1, \xi)\right|^2 d\xi.$$

By integrating with respect to x_2, we have

$$\int_0^{H+u(x_1)} |v(x_1,x_2)|^2 dx_2 \le \int_0^{H+u(x_1)} x_2\, dx_2 \int_0^{H+u(x_1)} \left|\frac{\partial v}{\partial x_2}(x_1,\xi)\right|^2 d\xi$$
$$= \frac{(H+u(x_1))^2}{2} \int_0^{H+u(x_1)} \left|\frac{\partial v}{\partial x_2}(x_1,\xi)\right|^2 d\xi \le \frac{d^2}{2} \int_0^{H+u(x_1)} \left|\frac{\partial v}{\partial x_2}(x_1,\xi)\right|^2 d\xi.$$

We now integrate with respect to x_1 and we obtain

$$\int_0^L \int_0^{H+u(x_1)} |v(x_1,x_2)|^2 dx_2\, dx_1 \le \frac{d^2}{2} \int_0^L \int_0^{H+u(x_1)} \left|\frac{\partial v}{\partial x_2}(x_1,\xi)\right|^2 d\xi\, dx_1$$

which implies

$$||v||^2_{0,\Omega_u} \le \frac{d^2}{2} \left\|\frac{\partial v}{\partial x_2}\right\|^2_{0,\Omega_u} \le \frac{d^2}{2} ||\nabla v||^2_{0,\Omega_u}.$$

We shall show that the inequality also holds for $v \in H^1(\Omega_u)$, $v = 0$ on Γ_u, by density. Let (v_k) be a sequence, $v_k \in C^1(\overline{\Omega}_u)$, $v_k = 0$ on Γ_u, such that $\lim_{k\to\infty} v_k = v$ for the norm $||\cdot||_{1,\Omega_u}$. We have to pass to the limit in the inequality $||v_k||^2_{0,\Omega_u} \le \frac{d^2}{2} ||\nabla v_k||^2_{0,\Omega_u}$. We have

$$\left| ||v_k||_{0,\Omega_u} - ||v||_{0,\Omega_u} \right| \le ||v_k - v||_{0,\Omega_u} \le ||v_k - v||_{1,\Omega_u},$$

thus $\lim_{k\to\infty} ||v_k||_{0,\Omega_u} = ||v||_{0,\Omega_u}$. Similarly, we have

$$\left| \left\|\frac{\partial v_k}{\partial x_i}\right\|_{0,\Omega_u} - \left\|\frac{\partial v}{\partial x_i}\right\|_{0,\Omega_u} \right| \le \left\|\frac{\partial v_k}{\partial x_i} - \frac{\partial v}{\partial x_i}\right\|_{0,\Omega_u} \le ||v_k - v||_{1,\Omega_u}$$

and $\lim_{k\to\infty} ||\nabla v_k||_{0,\Omega_u} = ||\nabla v||_{0,\Omega_u}$ giving [6.67]. □

If a domain Ω is regular, for example, Lipschitz, then $C^1(\overline{\Omega})$ and $\mathcal{D}(\overline{\Omega})$ are dense in $H^1(\Omega)$. The sets $C^1(\Omega)$ and $\mathcal{D}(\Omega)$ are also dense in $H^1_0(\Omega)$.

As $||v||^2_{1,\Omega_u} = ||v||^2_{0,\Omega_u} + ||\nabla v||^2_{0,\Omega_u}$, by using the Poincaré inequality, we obtain

$$||v||^2_{1,\Omega_u} \le \left(1 + \frac{d^2}{2}\right) ||\nabla v||^2_{0,\Omega_u}$$

which is equivalent to

$$\|v\|_{1,\Omega_u} \leq \sqrt{1+\frac{d^2}{2}}\,\|\nabla v\|_{0,\Omega_u}.$$

A demonstration of the Poincaré inequality for $H^1(\Omega)$ can be found in [DAU 88] (vol. 3, chap. IV, section 7.2, p. 422).

To obtain the uniform ellipticity [6.39], we shall first prove a simplified version of Korn's inequality, where we impose homogeneous Dirichlet boundary conditions over the entire boundary. For a demonstration of Korn's inequality for the case where we impose homogeneous Dirichlet boundary conditions over only a part of the boundary, consult [CIA 04] (Theorem 6.3-4, p. 292).

PROPOSITION 6.2.– *[Simplified version of Korn's inequality] We have*

$$\forall \mathbf{v} \in (H_0^1(\Omega_u))^2, \quad \int_{\Omega_u} \nabla\mathbf{v} : \nabla\mathbf{v}\, d\mathbf{x} \leq 2\int_{\Omega_u} \epsilon(\mathbf{v}) : \epsilon(\mathbf{v}) d\mathbf{x}. \qquad [6.68]$$

DEMONSTRATION 6.4.– We shall first demonstrate the inequality for $\mathbf{v} \in (\mathcal{D}(\Omega_u))^2$. We have

$$2\int_{\Omega_u} \epsilon(\mathbf{v}) : \epsilon(\mathbf{v}) d\mathbf{x} = \int_{\Omega_u} 2\left(\frac{\partial v_1}{\partial x_1}\right)^2 + 2\left(\frac{\partial v_2}{\partial x_2}\right)^2 + \left(\frac{\partial v_1}{\partial x_2}\right)^2 + \left(\frac{\partial v_2}{\partial x_1}\right)^2 d\mathbf{x}$$
$$+ 2\int_{\Omega_u} \frac{\partial v_1}{\partial x_2}\frac{\partial v_2}{\partial x_1} d\mathbf{x}.$$

We shall use integration by parts and the equality $\frac{\partial^2 v_2}{\partial x_1 \partial x_2} = \frac{\partial^2 v_2}{\partial x_2 \partial x_1}$ to obtain

$$\int_{\Omega_u} \frac{\partial v_1}{\partial x_2}\frac{\partial v_2}{\partial x_1} d\mathbf{x} = -\int_{\Omega_u} v_1 \frac{\partial^2 v_2}{\partial x_2 \partial x_1} d\mathbf{x} = -\int_{\Omega_u} v_1 \frac{\partial^2 v_2}{\partial x_1 \partial x_2} d\mathbf{x}$$
$$= \int_{\Omega_u} \frac{\partial v_1}{\partial x_1}\frac{\partial v_2}{\partial x_2} d\mathbf{x}.$$

We have the identity

$$(\nabla \cdot \mathbf{v})^2 = \left(\frac{\partial v_1}{\partial x_1}\right)^2 + \left(\frac{\partial v_2}{\partial x_2}\right)^2 + 2\frac{\partial v_1}{\partial x_2}\frac{\partial v_2}{\partial x_1}$$

and we obtain

$$2\int_{\Omega_u} \epsilon(\mathbf{v}) : \epsilon(\mathbf{v}) d\mathbf{x} = \int_{\Omega_u} \left(\frac{\partial v_1}{\partial x_1}\right)^2 + \left(\frac{\partial v_2}{\partial x_2}\right)^2 + \left(\frac{\partial v_1}{\partial x_2}\right)^2 + \left(\frac{\partial v_2}{\partial x_1}\right)^2 d\mathbf{x}$$

$$+ \int_{\Omega_u} (\nabla \cdot \mathbf{v})^2 d\mathbf{x} = \int_{\Omega_u} \nabla\mathbf{v} : \nabla\mathbf{v} d\mathbf{x} + \int_{\Omega_u} (\nabla \cdot \mathbf{v})^2 d\mathbf{x}$$

thus

$$2\int_{\Omega_u} \epsilon(\mathbf{v}) : \epsilon(\mathbf{v}) d\mathbf{x} \geq \int_{\Omega_u} \nabla\mathbf{v} : \nabla\mathbf{v} d\mathbf{x}.$$

As $\mathcal{D}(\Omega)$ is dense in $H_0^1(\Omega)$, we deduce [6.68]. □

We can now obtain the uniform ellipticity [6.39] for the case of homogeneous Dirichlet boundary conditions over the entire domain boundary. We have for $\mathbf{v} \in (H_0^1(\Omega_u))^2$

$$2\int_{\Omega_u} \epsilon(\mathbf{v}) : \epsilon(\mathbf{v}) d\mathbf{x} \geq \int_{\Omega_u} \nabla\mathbf{v} : \nabla\mathbf{v} d\mathbf{x} = ||\nabla v||_{0,\Omega_u}^2 \geq \frac{1}{1 + \frac{d^2}{2}} ||v||_{1,\Omega_u}^2$$

which implies [6.39] with $\alpha_F = \frac{\mu^F}{1+\frac{d^2}{2}}$.

We shall now study the constant in the trace theorem.

We use $\mathbb{R}_+^2 = \{(x_1, x_2);\ x_1 > 0\}$ to denote the upper half-space.

PROPOSITION 6.3.– *[trace] Let $\eta \geq \max_{x_1 \in [0,L]} |u'(x_1)|$, thus u is a Lipschitz function of constant η. Hence, there exists a constant $C(\eta) > 0$ independent of u, such that*

$$\forall v \in H^1(\Omega_u), \quad ||v||_{0,\Gamma_{in}} \leq C(\eta)\, ||v||_{1,\Omega_u}. \qquad [6.69]$$

DEMONSTRATION 6.5.– Let $\phi \in \mathcal{D}(\mathbb{R}^2)$. We have

$$\phi^2(x_1, x_2) = \int_0^{x_1} \frac{\partial}{\partial x_1}(\phi^2(\xi, x_2)) d\xi + \phi^2(0, x_2).$$

As ϕ has compact support, we deduce

$$\phi^2(0,x_2) = -\int_0^\infty \frac{\partial}{\partial x_1}(\phi^2(\xi,x_2))d\xi = -2\int_0^\infty \frac{\partial \phi}{\partial x_1}(\xi,x_2)\phi(\xi,x_2)d\xi$$

$$\leq \int_0^\infty \left(\frac{\partial \phi}{\partial x_1}(\xi,x_2)\right)^2 d\xi + \int_0^\infty \phi^2(\xi,x_2)d\xi.$$

By integrating with respect to x_2, we obtain

$$||\phi||^2_{0,Ox_2} = \int_{-\infty}^\infty \phi^2(0,x_2)dx_2 \leq \int_{-\infty}^\infty \int_0^\infty \left(\frac{\partial \phi}{\partial x_1}(\xi,x_2)\right)^2 d\xi dx_2$$

$$+ \int_{-\infty}^\infty \int_0^\infty \phi^2(\xi,x_2)d\xi dx_2 \leq ||\phi||^2_{1,\mathbb{R}^2_+}$$

which means that the trace constant is 1 for the half-space $\mathbb{R}^2_+$. By density, we obtain the same constant for the functions in $H^1(\mathbb{R}^2_+)$.

We shall use a uniform extension operator constructed in [CHE 75]. For $v \in H^1(\Omega_u)$, there exists $E(v) \in H^1(\mathbb{R}^2)$, such that $E(v) = v$ in Ω_u, $E(v) = 0$ outside a ball containing $\overline{\Omega}_u$ and

$$||E(v)||_{1,\mathbb{R}^2} \leq C(\eta)\, ||v||_{1,\Omega_u}$$

where the constant $C(\eta) > 0$ does not depend on Ω_u.

Finally, we obtain

$$||v||_{0,\Gamma_{in}} = ||E(v)||_{0,\Gamma_{in}} \leq ||E(v)||_{0,Ox_2} \leq ||E(v)||_{1,\mathbb{R}^2_+} \leq C(\eta)\, ||v||_{1,\Omega_u}. \qquad \square$$

6.6. Numerical tests

We have adapted the test from [NOB 01] (chapter 4, section 11). We have the geometric configuration presented in Figure 6.1, where the structure domain has length $L = 6$ cm and thickness $h = 0.1$ cm,

$\Gamma_D^S =]AB[\cup]CD[$ and $\Gamma_N^S =]DA[$. The reference domain for the fluid is a rectangle of length $L = 6\ cm$ and height $H = 1\ cm$.

The physical parameters for the structure are: Young's modulus $E = 3 \cdot 10^6 \frac{g}{cm \cdot s^2}$, Poisson's ratio $\nu^S = 0.3$, the mass density $\rho^S = 1.1 \frac{g}{cm^3}$ and the body forces $\mathbf{f}^S = (0, 0)^T$. The Lamé parameters are given by

$$\lambda^S = \frac{\nu^S E}{(1 - 2\nu^S)(1 + \nu^S)}, \quad \mu^S = \frac{E}{2(1 + \nu^S)}.$$

For the fluid, we use the physical parameters: the dynamic viscosity $\mu^F = 0.035 \frac{g}{cm \cdot s}$, the mass density $\rho^F = 1 \frac{g}{cm^3}$, the body forces $\mathbf{f}^F = (0, 0)^T$.

The boundary conditions are:

$$\begin{aligned}
\mathbf{U} &= 0, \text{ on } \Gamma_D^S \times]0, T[\\
\sigma^S \mathbf{N}^S &= 0, \text{ on } \Gamma_N^S \times]0, T[\\
\mathbf{v} &= 0, \text{ on } \Gamma_D^F \times]0, T[\\
\sigma^F \mathbf{n}^F &= \mathbf{h}_{in}^F, \text{ on } \Gamma_{in}^F \times]0, T[\\
\sigma^F \mathbf{n}^F &= \mathbf{h}_{out}^F, \text{ on } \Gamma_{out}^F \times]0, T[
\end{aligned}$$

where

$$\mathbf{h}_{in}^F(\mathbf{x}, t) = \begin{cases} \left(10^3(1 - \cos(2\pi t/0.025)),\ 0\right), & 0 \le t \le 0.025 \\ (0,\ 0), & 0.025 < t \end{cases}$$

and $\mathbf{h}_{out}^F = (0,\ 0)$.

We study the fluid–structure interaction in the time interval $[0, T = 0.1]$ with zero initial conditions.

6.6.1. *Implicit domain calculation*

We use triangular finite elements: $\mathbb{P}_1 + b$ for the fluid velocity, and $\mathbb{P}_1$ for the fluid pressure, the structure displacement and the ALE map.

We have used the following numerical parameters: the time step-size of $\Delta t = 0.001\ s$, $N = 100$ steps in time, the relaxation factor $\omega = 0.03$ and for the stopping test $tol = 10^{-4}$.

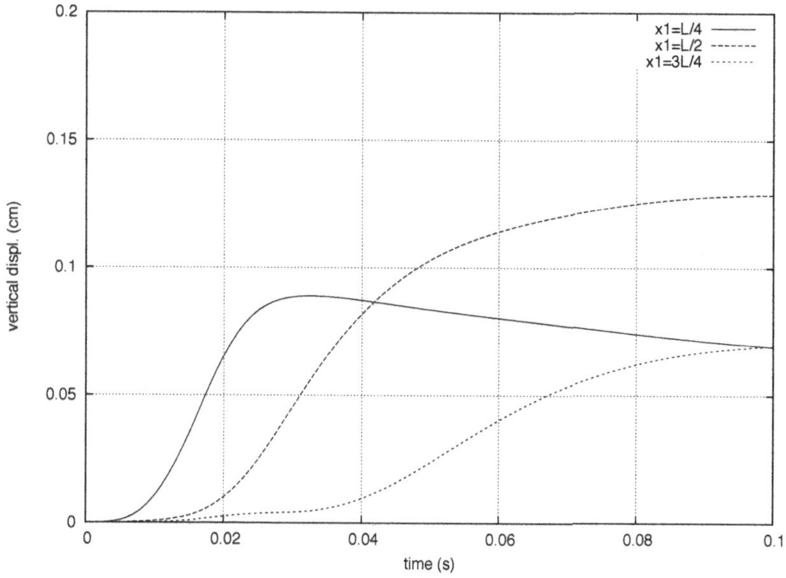

Figure 6.2. *The vertical displacement of three points at the interface for* $x_1 = L/4$, $x_1 = L/2$ *and* $x_1 = 3L/4$

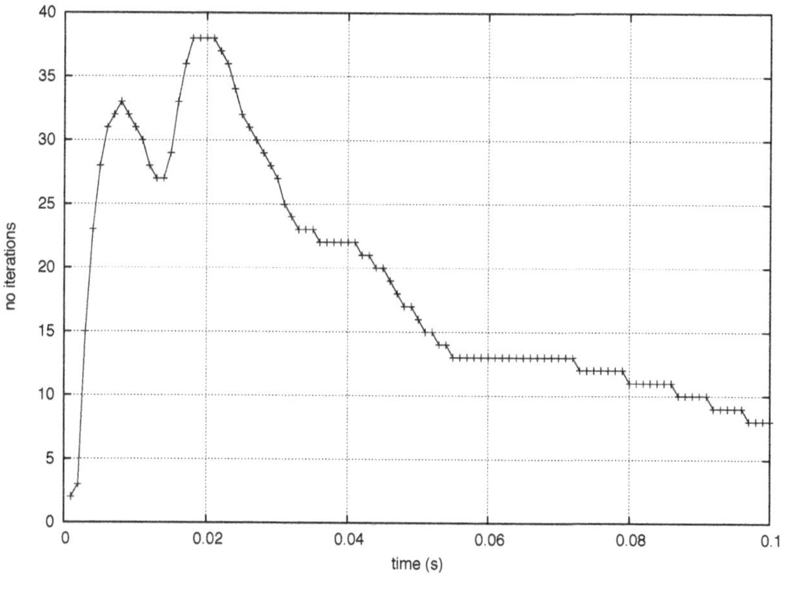

Figure 6.3. *The time evolution of the number of fixed point iterations, at each step in time*

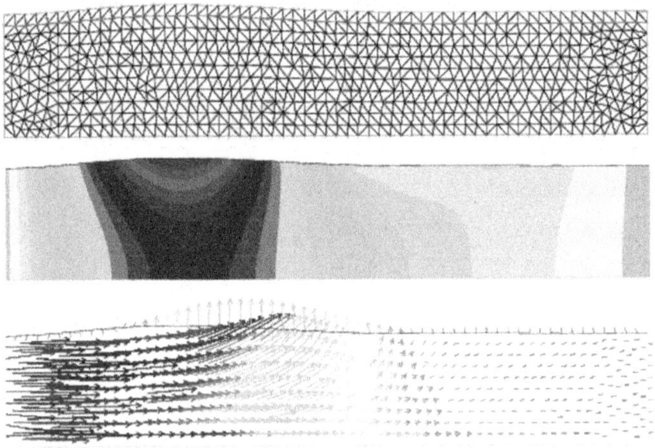

Figure 6.4. *The numerical solution at time* $t = 0.025$*: the structure and fluid meshes (top), the pressure (middle) and the fluid velocity (bottom). For a color version of this figure, see www.iste.co.uk/murea/schemes.zip*

Appendix

Functional Analysis Tools

A.1. Sobolev spaces

For the theory of Sobolev spaces, consult [ADA 75], [BRÉ 05], [DAU 88, vol. 3, Chap. IV], [BAR 98].

Let $\Omega \subset \mathbb{R}^2$ be an open, non-empty set. We denote its boundary as $\partial\Omega$ and its closure as $\overline{\Omega}$.

The scalar product and the norm of the space $L^2(\Omega)$ are denoted as:

$$(v, w)_{0,\Omega} = \int_\Omega v(\mathbf{x})w(\mathbf{x})\, d\mathbf{x}, \quad \|v\|_{0,\Omega} = \sqrt{(v, v)_{0,\Omega}}.$$

The Sobolev space of order one on Ω is defined by:

$$H^1(\Omega) = \left\{ w \in L^2(\Omega);\ \frac{\partial w}{\partial x_i} \in L^2(\Omega),\ i = 1, 2 \right\}$$

with the derivative in the distributional sense. We note the scalar product and the norm as:

$$(v, w)_{1,\Omega} = \int_\Omega v(\mathbf{x})w(\mathbf{x})\, d\mathbf{x} + \int_\Omega \nabla v \cdot \nabla w\, d\mathbf{x}, \quad \|v\|_{1,\Omega} = \sqrt{(v, v)_{1,\Omega}}$$

and $H^1(\Omega)$ is a Hilbert space for this scalar product.

We will adopt the following notation for $m \in \mathbb{N}^*$:

$$C^m(\Omega) = \left\{w : \Omega \to \mathbb{R};\ \forall \alpha,\ |\alpha| \leq m,\ D^\alpha w \in C^0(\Omega)\right\},$$

where $\alpha = (\alpha_1, \alpha_2) \in \mathbb{N} \times \mathbb{N}$ is a multi-index with $|\alpha| = \alpha_1 + \alpha_2$ and $D^\alpha w = \frac{\partial^{|\alpha|} w}{\partial x_1^{\alpha_1} \partial x_2^{\alpha_2}}$. We have the inclusion $C^1(\Omega) \subset H^1(\Omega)$. We can see the space $H^1(\Omega)$ as the analogue of $C^1(\Omega)$, where the classical derivative has been replaced by the derivative in the distributional sense. We label $\mathcal{D}(\Omega)$ as the space of infinitely differentiable functions with compact support in Ω.

If the open set Ω is bounded, we note for $m \in \mathbb{N}^*$:

$$C^m(\overline{\Omega}) = \left\{w \in C^m(\Omega);\ \forall \alpha,\ |\alpha| \leq m,\ \exists f^\alpha \in C^0(\overline{\Omega}), D^\alpha w = f^\alpha \text{ in } \Omega\right\},$$

in other words, the derivatives $D^\alpha w : \Omega \to \mathbb{R}$ can be extended by continuity onto the boundary $\partial\Omega$. We use $\mathcal{D}(\overline{\Omega})$ to denote the space of functions $w : \overline{\Omega} \to \mathbb{R}$ with compact support in $\overline{\Omega}$ and $w \in C^m(\overline{\Omega})$ for all $m \in \mathbb{N}$.

We will define $H_0^1(\Omega)$ as the closing of $\mathcal{D}(\Omega)$ in $H^1(\Omega)$ for the norm $\|\cdot\|_{1,\Omega}$.

In order to introduce the trace theorem, we require supplementary hypotheses regarding the regularity of the boundary $\partial\Omega$.

DEFINITION A.1.– *[see [CHE 75, HEN 05]] Let r, a, L be three positive numbers and Ω be an open, bounded set in $\mathbb{R}^2$. We say that its boundary $\partial\Omega$ is Lipschitz if, for every $\mathbf{x}_0 \in \partial\Omega$, there exists a system of Cartesian coordinates $\{x_1, x_2\}$ with origin $\mathbf{x}_0$ and a Lipschitz function $\zeta : (-r, r) \to (-a, a)$ with constant L, such that $\zeta(0) = 0$,*

$$\partial\Omega \cap P(\mathbf{x}_0) = \{(x_1, \zeta(x_1)),\ x_1 \in (-r, r)\},$$

$$\Omega \cap P(\mathbf{x}_0) = \{(x_1, x_2),\ x_1 \in (-r, r),\ \zeta(x_1) < x_2 < a\},$$

where $P(\mathbf{x}_0) = (-r, r) \times (-a, a)$.

For an open, bounded domain with Lipschitz boundary Ω, we can define the space $L^2(\partial\Omega)$ equipped with the scalar product and the norm as:

$$(v, w)_{0,\partial\Omega} = \int_{\partial\Omega} v(\mathbf{s}) w(\mathbf{s})\, ds, \quad \|v\|_{0,\partial\Omega} = \sqrt{(v, v)_{0,\partial\Omega}}.$$

THEOREM A.1.– *[trace] If* Ω *is an open, bounded set with a Lipschitz boundary, the map* $\gamma_0 : \mathcal{D}(\overline{\Omega}) \to C^0(\partial\Omega)$ *extends by continuity to a linear and continuous map of* $H^1(\Omega)$ *in* $L^2(\partial\Omega)$*, still denoted as* γ_0.

We have:

$$H_0^1(\Omega) = \left\{ w \in H^1(\Omega);\ \gamma_0(w) = 0 \text{ in } L^2(\partial\Omega) \right\}.$$

The map $\gamma_0 : H^1(\Omega) \to L^2(\partial\Omega)$ is not surjective, we have $\gamma_0\left(H^1(\Omega)\right) = H^{1/2}(\partial\Omega)$, where $H^{1/2}(\partial\Omega)$ is a proper vector sub-space of $L^2(\partial\Omega)$. In [DAU 88], vol. 3, Chap. IV, p. 943, the Sobolev space $H^{1/2}(\partial\Omega)$ is defined by charts and a partition of unity on $\partial\Omega$. It is a Hilbert space with the scalar product denoted as $(v, w)_{1/2,\partial\Omega}$. The associated norm $\|g\|_{1/2,\partial\Omega} = \sqrt{(g, g)_{1/2,\partial\Omega}}$ is equivalent to the norm on $H^{1/2}(\partial\Omega)$:

$$\|g\|_* = \inf\{\|w\|_{1,\Omega};\ w \in H^1(\Omega),\ \gamma_0(w) = g\}.$$

In a Hilbert space W with the scalar product $(\cdot, \cdot)_W$ and the associated norm $\|w\|_W = \sqrt{(w, w)_W}$, we have the inequality:

$$\forall v, w \in W, \quad |(v, w)_W| \leq \|v\|_W \|w\|_W.$$

For the space $L^2(\Omega)$, where $\Omega \subset \mathbb{R}^2$ is open, non-empty and not necessarily bounded, we have the Cauchy-Schwarz inequality:

$$\forall v, w \in L^2(\Omega) \quad |(v, w)_{0,\Omega}| \leq \|v\|_{0,\Omega} \|w\|_{0,\Omega}. \qquad \text{[A.1]}$$

In $\mathbb{R}^n$, the Cauchy-Schwarz inequality gives:

$$\left| \sum_{k=1}^{n} \alpha_k \beta_k \right| \leq \left(\sum_{k=1}^{n} \alpha_k^2 \right)^{1/2} \left(\sum_{k=1}^{n} \beta_k^2 \right)^{1/2} \qquad \text{[A.2]}$$

where $\alpha_k, \beta_k \in \mathbb{R}$, $k = 1, \ldots, n$. This inequality is also called the Cauchy-Bunyakovsky-Schwarz inequality.

THEOREM A.2.– *[Hölder's inequality] Let* $\Omega \subset \mathbb{R}^2$ *be an open, non-empty and not necessarily bounded set, and* $f_1, f_2, \ldots, f_k$ *functions such that* $f_i \in L^{p_i}(\Omega)$,

$p_i \geq 1$, $1 \leq i \leq k$ *with* $\frac{1}{p} = \frac{1}{p_1} + \cdots + \frac{1}{p_k} \leq 1$. *Then, the product* $f = f_1 f_2 \dots f_k \in L^p(\Omega)$ *and*

$$\|f\|_{L^p(\Omega)} \leq \|f_1\|_{L^{p_1}(\Omega)} \cdots \|f_k\|_{L^{p_k}(\Omega)}\,. \qquad [A.3]$$

DEFINITION A.2.– *A band in* $\mathbb{R}^2$ *is a set obtained by translation and rotation of the set* $\mathbb{R}\times]0, d[$, *where* $0 < d < \infty$.

THEOREM A.3.– *[Poincaré's inequality] If* Ω *is an open, non-empty set, included in a band, then:*

$$\exists C_P(\Omega) > 0,\ \forall w \in H_0^1(\Omega),\ \|w\|_{0,\Omega} \leq C_P(\Omega)\,\|\nabla w\|_{0,\Omega}\,. \qquad [A.4]$$

THEOREM A.4.– *[Korn's inequality] If* Ω *is an open, non-empty, bounded, connected set with a Lipschitz boundary, then:*

$$\exists C_K(\Omega) > 0,\ \forall \mathbf{w} \in (H_0^1(\Omega))^2, \quad \|\mathbf{w}\|_{1,\Omega} \leq C_K(\Omega)\,\|\boldsymbol{\epsilon}(\mathbf{w})\|_{0,\Omega} \qquad [A.5]$$

where $\boldsymbol{\epsilon}(\mathbf{w}) = \frac{1}{2}\left(\nabla\mathbf{v} + (\nabla\mathbf{v})^T\right)$.

DEFINITION A.3.– *Let* $\Omega \subset \mathbb{R}^2$ *be an open, non-empty, bounded, connected set with a Lipschitz boundary. We use* $\mathbf{n}(\mathbf{x}) = (n_1(\mathbf{x}), n_2(\mathbf{x}))$ *to denote the unit vector, normal to the boundary at point* $\mathbf{x} \in \partial\Omega$, *pointing out of* Ω. *We accept the following result: the vector* $\mathbf{n}(\mathbf{x})$ *is well-defined almost everywhere on the Lipschitz boundary* $\partial\Omega$.

THEOREM A.5.– *[Green's formula] Let* $\Omega \subset \mathbb{R}^2$ *be an open, non-empty, bounded, connected set with a Lipschitz boundary. If* $u, v \in H^1(\Omega)$, *then, for* $i = 1, 2$, *we have:*

$$\int_\Omega u(\mathbf{x})\frac{\partial v(\mathbf{x})}{\partial x_i}d\mathbf{x} = \int_{\partial\Omega} u(\mathbf{s})v(\mathbf{s})n_i(\mathbf{s})d\mathbf{s} - \int_\Omega \frac{\partial u(\mathbf{x})}{\partial x_i}v(\mathbf{x})d\mathbf{x}.$$

The gradient of a vector-valued function $\mathbf{w} = (w_1, w_2) \in \left(H^1(\Omega)\right)^2$ is denoted by:

$$\nabla\mathbf{w} = \begin{pmatrix} \dfrac{\partial w_1}{\partial x_1} & \dfrac{\partial w_1}{\partial x_2} \\ \dfrac{\partial w_2}{\partial x_1} & \dfrac{\partial w_2}{\partial x_2} \end{pmatrix}.$$

The divergence operator acting on a vector-valued function $\mathbf{w} = (w_1, w_2) \in \left(H^1(\Omega)\right)^2$ and on a tensor $\boldsymbol{\sigma} = \left(\sigma_{ij}\right)_{1\leq i,j\leq 2}$, $\boldsymbol{\sigma} \in \left(H^1(\Omega)\right)^4$ are denoted by:

$$\nabla \cdot \mathbf{w} = \frac{\partial w_1}{\partial x_1} + \frac{\partial w_2}{\partial x_2}, \qquad \nabla \cdot \boldsymbol{\sigma} = \begin{pmatrix} \dfrac{\partial \sigma_{11}}{\partial x_1} + \dfrac{\partial \sigma_{12}}{\partial x_2} \\ \dfrac{\partial \sigma_{21}}{\partial x_1} + \dfrac{\partial \sigma_{22}}{\partial x_2} \end{pmatrix}.$$

If $\boldsymbol{\sigma} = \left(\sigma_{ij}\right)_{1\leq i,j\leq 2}$ and $\boldsymbol{\tau} = \left(\tau_{ij}\right)_{1\leq i,j\leq 2}$ are two tensors, we note:

$$\boldsymbol{\sigma} : \boldsymbol{\tau} = \sum_{i=1}^{2} \sum_{j=1}^{2} \sigma_{ij}\tau_{ij}.$$

From Green's formula, we have, for $\mathbf{w} \in \left(H^1(\Omega)\right)^2$ and $q \in H^1(\Omega)$,

$$\int_{\Omega} (\nabla \cdot \mathbf{w}) q \, d\mathbf{x} = \int_{\partial\Omega} (\mathbf{w} \cdot \mathbf{n}) q \, ds - \int_{\Omega} \mathbf{w} \cdot \nabla q \, d\mathbf{x},$$

and, for $\boldsymbol{\sigma} \in \left(H^1(\Omega)\right)^4$ and $\mathbf{w} \in \left(H^1(\Omega)\right)^2$,

$$\int_{\Omega} (\nabla \cdot \boldsymbol{\sigma}) \cdot \mathbf{w} \, d\mathbf{x} = \int_{\partial\Omega} (\boldsymbol{\sigma}\mathbf{n}) \cdot \mathbf{w} \, ds - \int_{\Omega} \boldsymbol{\sigma} : \nabla \mathbf{w} \, d\mathbf{x}.$$

A.2. Closed, surjective operators

We now recall a result that characterizes closed, surjective operators (see [BRÉ 05]).

Let E and F be two Banach spaces and A be a linear, non-bounded operator from E to F:

$$A : D(A) \subset E \longrightarrow F,$$

where $D(A)$ is the domain of A.

We call the vector sub-space of $E \times F$, which is defined by: $G(A) = \bigcup_{u \in D(A)} [u, Au] \subset E \times F$, the *graph* of A.

We call the vector sub-space of F, which is defined by: $R(A) = \bigcup_{u \in D(A)} Au \subset F$, the *image* of A.

We call the vector sub-space of E, which is defined by: $N(A) = \{u \in D(A), Au = 0\} \subset E$, the *kernel* of A.

DEFINITION A.4.– *We say that an operator A is closed if $G(A)$ is closed in $E \times F$.*

We denote the dual space of E with $E' = \mathcal{L}(E, \mathbb{R})$ and the duality pairing between E' and E by $\langle \cdot, \cdot \rangle_{E',E}$. If $f \in E'$, then:

$$\|f\|_{E'} = \sup_{u \in E, u \neq 0} \frac{|\langle f, u \rangle_{E',E}|}{\|u\|_E}.$$

Let:

$$D(A^*) \overset{déf}{=} \{v \in F'; \exists c \geq 0, \forall u \in D(A), |\langle v, Au \rangle_{F',F}| \leq c \, \|u\|_E\}.$$

The adjoint operator of A, denoted as $A^* : D(A^*) \subset F' \to E'$, satisfies:

$$\langle v, Au \rangle_{F',F} = \langle A^* v, u \rangle_{E',E}, \quad \forall u \in D(A), \forall v \in D(A^*).$$

We can now state the characteristic result for a closed operator. A demonstration can be found in [BRÉ 05] (Theorem II.18, p. 29).

THEOREM A.6.– *Let $A : D(A) \subset E \longrightarrow F$ be a linear, non-bounded closed operator, with $\overline{D(A)} = E$. The following properties are equivalent:*

i) $R(A)$ is closed;

ii) $R(A^*)$ is closed;

iii) $R(A) = (N(A^*))^\circ \overset{def}{=} \{y \in F;\ \langle f, y \rangle_{F',F} = 0,\ \forall f \in N(A^*)\}$;

iv) $R(A^*) = (N(A))^\circ \overset{def}{=} \{f \in E';\ \langle f, x \rangle_{E',E} = 0,\ \forall x \in N(A)\}$.

We can state the characteristic result for a surjective operator. A demonstration can be found in [BRÉ 05] (Theorem II.19, p. 30).

THEOREM A.7.– *Let* $A : D(A) \subset E \longrightarrow F$ *be a linear, non-bounded, closed operator with* $\overline{D(A)} = E$*, and* $A^* : D(A^*) \subset F' \rightarrow E'$ *be the adjoint of* A*. The following properties are equivalent:*

i) A *is surjective*

ii) There exists a constant $C > 0$ *such that*

$$\|v\|_{F'} \leq C \left\|A^* v\right\|_{E'}, \quad \forall v \in D(A^*)$$

iii) $N(A^*) = \{0\}$ *and* $R(A^*)$ *is closed.*

DEFINITION A.5.– *Let* W *be a Hilbert space. Every element* $u \in W$ *defines an element* $f_u \in W'$ *according to*

$$\langle f_u, w\rangle_{W',W} = (u, w)_W, \quad \forall w \in W.$$

We have $\|f_u\|_{W'} = \|u\|_W$*. We can define a map* $j : W \rightarrow W'$ *by* $j(u) = f_u$ *that is linear, continuous and injective that we call the canonical injection from* W *to* W'*.*

The following result shows that, for a Hilbert space, the canonical injection is surjective.

THEOREM A.8.– *[Riesz-Fréchet representation theorem] Let* W *be a Hilbert space and* $f \in W'$*. Hence, there exists a unique* $w_f \in W$ *such that:*

$$\langle f, w\rangle_{W',W} = (w_f, w)_W, \quad \forall w \in W.$$

Furthermore, $\|f\|_{W'} = \left\|w_f\right\|_W$*.*

A demonstration of the above result can be found in [BRÉ 05, p. 81].

THEOREM A.9.– *[Lax-Milgram] Let* V *be a Hilbert space. Let* $a : V \times V \rightarrow \mathbb{R}$ *be a bilinear, continuous and* V*-elliptic map, i.e.*

$$\exists \alpha > 0, \ \forall v \in V, \quad \alpha \|v\|_V^2 \leq a(v, v) \qquad [A.6]$$

and $L \in V'$. Hence, there exists a unique $u \in V$ such that:

$$\forall v \in V, \quad a(u, v) = \langle L, v \rangle_{V',V}. \qquad \text{[A.7]}$$

We have $\|u\|_V \leq \frac{1}{\alpha} \|L\|_{V'}$.

A demonstration of the above result can be found in [BRÉ 05, p. 84].

Bibliography

[ADA 75] ADAMS R., *Sobolev Spaces*, Academic Press, New York, 1975.

[ARN 84] ARNOLD D.N., BREZZI F., FORTIN M., "A stable finite element for the Stokes equations", *Calcolo*, vol. 21, no. 4, pp. 337–344 (1985), 1984.

[BAB 73] BABUŠKA I., "The finite element method with Lagrangian multipliers", *Numerische Mathematik*, vol. 20, pp. 179–192, 1973.

[BÄN 00] BÄNSCH E., HÖHN B., "Numerical treatment of the Navier-Stokes equations with slip boundary condition", *SIAM Journal on Scientific Computing*, vol. 21, no. 6, pp. 2144–2162, 2000.

[BAR 98] BARBU V., *Partial Differential Equations and Boundary Value Problems*, Springer, Netherlands, 1998.

[BER 79] BERCOVIER M., ENGELMAN M., "A finite element for the numerical solution of viscous incompressible flows", *Journal of Computational Physics*, vol. 30, no. 2, pp. 181–201, 1979.

[BOF 13] BOFFI D., BREZZI F., FORTIN M., *Mixed Finite Element Methods and Applications*, Springer, Heidelberg, 2013.

[BRA 01] BRAESS D., *Finite Elements. Theory, Fast Solvers, and Applications in Solid Mechanics*, 2nd ed., Cambridge University Press, Cambridge, 2001.

[BRÉ 05] BRÉZIS H., *Analyse Fonctionnelle: Théorie et Applications*, Dunod, Paris, 2005.

[BRE 74] BREZZI F., "On the existence, uniqueness and approximation of saddle-point problems arising from Lagrangian multipliers", *Revue française d'automatique, informatique, recherche opérationnelle. Analyse numérique*, vol. 8, no. R-2, pp. 129–151, 1974.

[CAU 05] CAUSIN P., GERBEAU J.F., NOBILE F., "Added-mass effect in the design of partitioned algorithms for fluid-structure problems", *Computer Methods in Applied Mechanics and Engineering*, vol. 194, nos. 42–44, pp. 4506–4527, 2005.

[CHE 75] CHENAIS D., "On the existence of a solution in a domain identification problem", *Journal of Mathematical Analysis and Applications*, vol. 52, no. 2, pp. 189–219, 1975.

[CIA 04] CIARLET P.G., *Mathematical Elasticity: Three Dimensional Elasticity*, vol. 1, Elsevier, Netherlands, 2004.

[CLÉ 75] CLÉMENT P., "Approximation by finite element functions using local regularization", *Revue française d'automatique, informatique, recherche opérationnelle. Analyse numérique*, vol. 9, nos. R-2, pp. 77–84, 1975.

[DAU 88] DAUTRAY R., LIONS J.-L., *Analyse mathématique et calcul numérique pour les sciences et les techniques*, vol. 3, 4, 9, Masson, Paris, 1988.

[ERN 04] ERN A., GUERMOND J.-L., *Theory and Practice of Finite Elements*, Springer-Verlag, New York, 2004.

[FER 09] FERNÁNDEZ M.A., GERBEAU J.-F., "Algorithms for fluid-structure interaction problems", in FORMAGGIA L., QUARTERONI A., VENEZIANI A. (eds), *Cardiovascular Mathematics. Modeling Simulation and Applications*, vol. 1, Springer Italia, Milan, 2009.

[FOR 77] FORTIN M., "An analysis of the convergence of mixed finite element methods", *Revue française d'automatique, informatique, recherche opérationnelle. Analyse numérique*, vol. 11, no. 4, pp. 341–354, 1977.

[FOR 82] FORTIN M., GLOWINSKI R., *Méthodes de lagrangien augmenté*, Gauthier-Villars, Paris, 1982.

[GIR 86] GIRAULT V., RAVIART P.-A., *Finite Element Approximation of the Navier-Stokes Equations. Theory and Algorithms*, Springer-Verlag, Berlin, 1986.

[HEC 12] HECHT F., "New development in freefem++", *Journal of Numerical Mathematics*, vol. 20, no. 3–4, pp. 251–265, 2012.

[HEN 05] HENROT A., PIERRE M., *Variation et optimisation de formes. Une analyse geométrique*, Springer, 2005.

[KEL 76] KELLOGG R.B., OSBORN J.E., "A regularity result for the Stokes problem in a convex polygon", *Journal of Functional Analysis*, vol. 21, no. 4, pp. 397–431, 1976.

[NOB 01] NOBILE F., Numerical approximation of fluid-structure interaction problems, with application to haemodynamics, PhD Thesis, Ecole Polytechnique Féderale de Lausanne, Switzerland, 2001.

[PIR 89] PIRONNEAU O., *Finite Element Methods for Fluids*, John Wiley & Sons, Chichester, 1989.

[QUA 04] QUARTERONI A., FORMAGGIA L., "Mathematical modelling and numerical simulation of the cardiovascular system", in AYACHE N. (ed.), *Modeling of Living Systems*, North-Holland, Amsterdam, 2004.

[RAV 98] RAVIART P.-A., THOMAS J.-M., *Introduction l'analyse numérique des Equations aux dèrivèes partielles*, Dunod, Paris, 1998.

[SCH 96] SCHAEFER M., TUREK S., "Benchmark computations of laminar flow around cylinder", in HIRSCHEL E. (ed.), *Flow Simulation with High-Performance Computers II*, Vieweg, 1996.

[SCH 13] SCHEID J.-F., Analyse numérique des Equations de Navier-Stokes, Thesis, University of Lorraine, 2013.

[SY 08] SY S., MUREA C.M., "A stable time advancing scheme for solving fluid-stucture interaction problem at small structural displacements", *Computer Methods in Applied Mechanics and Engineering*, vol. 198, no. 2, pp. 210–222, 2008.

[TEM 01a] TEMAM R., MIRANVILLE A., *Mathematical Modeling in Continuum Mechanics*, Cambridge University Press, Cambridge, 2001.

[TEM 01b] TEMAM R., *Navier-Stokes Equations. Theory and Numerical Analysis*, AMS Chelsea Publishing, Providence, 2001.

[THO 77] THOMAS J.-M., Sur l'analyse numérique des méthodes d'Eléments finis hybrides et mixtes, Thesis, Pierre and Marie Cuire University, Paris, 1977.

[VER 87] VERFÜRTH R., "Finite element approximation of incompressible Navier-Stokes equations with slip boundary condition", *Numerische Mathematik*, vol. 50, no. 6, pp. 697–721, 1987.

Index

Printed in the United States
By Bookmasters